P9-CDH-581

Fine WoodWorking *on* Faceplate Turning

Fine WoodWorking *on* Faceplate Turning

42 articles selected by the editors of *Fine Woodworking* magazine

The Taunton Press

Cover photo by Wendell Smith

© 1987 by The Taunton Press, Inc.
All rights reserved

First printing: January 1987
International Standard Book Number: 0-918804-72-8
Library of Congress Catalog Card Number: 86-51289
Printed in the United States of America

A FINE WOODWORKING Book

FINE WOODWORKING® is a trademark of The Taunton Press, Inc.,
registered in the U.S. Patent and Trademark Office.

The Taunton Press, Inc.
63 South Main Street
Box 355
Newtown, Connecticut 06470

Contents

Introduction

Most woodworking machines work in the same way: the motor rotates the tool, while the operator holds and guides the workpiece. However, the lathe (which was the first woodworking machine) works the other way around: the machine rotates the workpiece, while the operator holds and guides the tool.

Virtually any lump of wood can be buttoned onto the business end of a motor and made to whirl around. And just about any bit of metal can be stuck into the whirling wood. The chips will fly everywhere, the wood will get smaller, the tool duller. Do this with some little skill and care, and you'll make a plate, or a bowl. This is faceplate turning.

In 42 articles reprinted from the first ten years of *Fine Woodworking* magazine, authors who are also craftsmen show the incredible diversity that can result from this simple process. (A companion volume, *Spindle Turning*, discusses work turned between centers.) Here is advice on how to harvest your own turning wood, how to prepare glued-up turning blanks, and how to use the modern bowl-turning gouge. There's also a wealth of turning ideas and projects.

John Kelsey, editor

The Turned Bowl

The end of infancy for a craft reborn

by John Kelsey

For more than 3,000 years, sturdy bowls and plates of turned wood were among the most ordinary kitchen utensils. But during the past 100 years, the useful wooden bowl has been supplanted by mass-produced ware of ceramic, glass, metal and plastic.

Forty years ago, James Prestini added a dimension to traditional woodturning: the delicate, decorative wooden bowl. During the past ten years this new craft of the turned bowl has blossomed and matured.

This contemporary flowering became publicly apparent in September 1981 in Philadelphia, when the Turned Objects Exhibition opened in conjunction with the Tenth Woodturning Symposium. The exhibition consists of 100 contemporary wooden turnings, selected by three jurors from about 1,500 entries. It will be traveling around the country during the next few years, provided sponsors can be found. The symposium was a long weekend of technical demonstrations and aesthetic argument among 30 expert turners, with 150 other turners, both amateur and professional, looking on and joining in. At the same time it was the summation of a five-year adventure, and the end of infancy for this craft reborn.

Along with a whole crowd of woodturners, I spent a day of that September symposium in a gallery filled by those 100 turned objects. Many of the 60 makers represented were there, as was juror David Ellsworth and symposium organizer Albert LeCoff.

We'd gathered to talk about this craft and its evolution, and to argue a little too. We tried to get beyond mutual admiration (hey, what a beautiful piece of wood) and on to what else might be said, not about turning tools and techniques but about design and aesthetics. Many of the turners in the room were having the first chance of their craft lives to

The Art of Prestini

"...It is hard to make place for Prestini among conventional craftsmen, and his place among artists would be exceptional and marginal. Yet his place is secure as a maker of beautiful, pure shapes.... This feat has been Prestini's, to suggest within the limits of simple craft the human pathos of art and the clean, bold certainties of science. He has made grand things that are not overwhelming, beautiful things that are not personal unveilings, and simple things that do not urge usefulness to excuse their simplicity. They are not precisely works of science or art, craft or convenience. Yet in their restraint and in their superb, direct assurance they touch our scope and potentialities, our limits and desires...."

—Edgar Kaufmann, Jr.: "Prestini's Art in Wood," Lake Forest Ill., Pochahontas Press, 1950. Mexican mahogany baseless bowl, 5¾ in. by 15¾ in., Museum of Modern Art Collection. Photo: © Barbara Morgan.

discuss their work with their peers, to hear artisans whose work they'd admired talk about ideas and values. I've been guided by that day of talk in choosing the turnings from the show to include in the second half of this article. But first, some highlights from 3,000 years of woodturning history.

The oldest turnings—The first archaeological fragments that seem to have been turned from wood are about 2,600 years old. They've been dug up in northern Italy and in Asia Minor. Archaeologists know a lot more about pottery than about turned wood, because potsherds don't decay the way woodchips do. The earliest turned wooden object to have survived intact is a bowl from a burial mound in Bavaria, about 600 B.C. As the sketch at right shows, the turner left toolmarks in a decorative ridged pattern and turned a ring free of the bowl's stem. Such sophistication indicates the lathe was already well known, although there's no direct evidence of early lathes until about 300 B.C. Writes the historian Robert S. Woodbury, "It seems quite clear that the lathe was in use as early as the eighth century B.C., probably as early as 1,000 B.C., and possibly even in 1,200 B.C. The place of its origin cannot be established, or even whether it had a single origin. . . possibly it was discovered independently by the Etruscans, the Celts and in Crimea." *(Studies in the History of Machine Tools,* Cambridge: The MIT Press, 1972.)

A few examples of treen survive from 17th-century America, along with thousands of objects from the 18th and early 19th centuries. The most durable woodenware to come down to us was turned from ash burl. Burl is more or less bowl-shaped as it comes off the tree, but man-powered lathes are not easy to work. I suspect it took about as much trouble to cut the burl and turn a bowl as it took to dig some clay, throw pots and fire them. Burl kitchenware is sturdier than pottery, until it gets on the wrong side of the moisture exchange. Then it's liable to crack wide open.

By 1850, industrial methods of forming clay, glass and metal had made turned wooden kitchenware all but obsolete. The craft of the turned utensil lingered in a few forms for which wood is particularly suitable (breadboards and rolling pins), as a hobby for grandpa, and in high-school shop class. Remember shop class? Lamp bases like sawed-off newel posts, nut bowls the shape of doggie dishes with green felt glued to their bottoms, honey-dippers drooling varnish?

The lathe (along with the bow drill) is man's oldest machine tool. It's probably the safest (and certainly the quickest) way to make raw wood into finished things. Turning ornamental intricacies on the Holtzapffel lathe is highly jigged and thus more akin to metalworking than to woodworking; but for that Victorian excess, traditional woodturning is staunchly utilitarian, not the place for startling beauty, nor for innovation. Then during the 1940s, James Prestini conceived the delicately thin, perfectly shaped, turned wooden bowl.

The pioneer—James Prestini, at 73, is a sculptor of iron and steel. Among many other things, he's been a professor of design, a research engineer, a mathematician and a metalworker. For 20 years, from 1933 until 1953, he was a woodturner. Although he turned and sold hundreds and hundreds of bowls and plates, it was always a hobby for him, not a full-time profession.

Prestini taught himself the woodturning trade by doing it. He entered the 3,000-year tradition, but he was not of that

Merryll Saylan (center) explains her turned sculpture, Jelly Doughnut, *during the Tenth Woodturning Symposium. The doughnut was assembled from eight mitered segments, seven of them of poplar and the eighth of red acrylic plastic.*

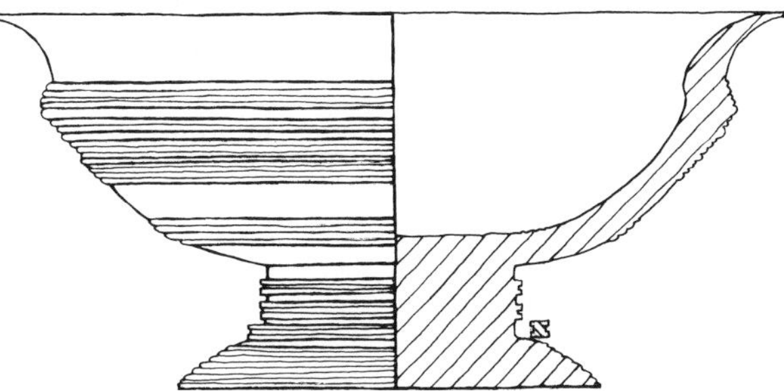

The oldest complete turning known to archaeology is this Celtic ceremonial bowl from the sixth century B.C., (above). Note the tooled decoration and the free-turned ring about its foot. Drawing adapted from Woodbury, Studies in the History of Machine Tools *(Cambridge: The MIT Press, 1972). Below, some traditional forms for utilitarian turned bowls, from Seale,* Practical Designs for Wood Turning *(London: 1964, reissued New York: Sterling, 1979.)*

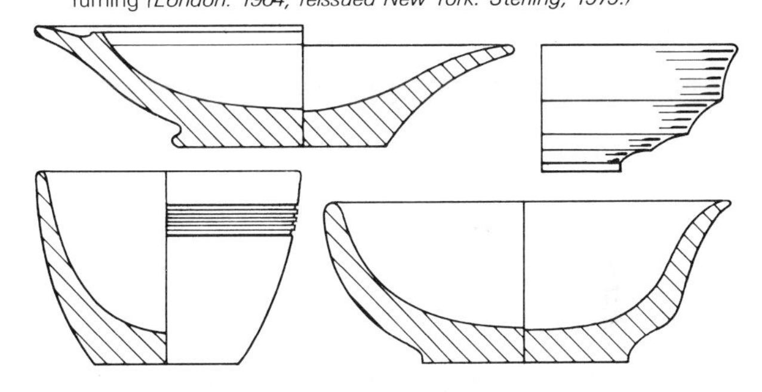

James Prestini in his library, with two of his sculptures in nickel-plated steel: Construction No. 161, 24 in. high (left), and Construction No. 286, 10½ in. dia. Photo: Jonathan Reichek/Catalyst.

From *Fine Woodworking* magazine (January 1982) 32:54-60

Decorative bowls by Bob Stocksdale, from his one-man exhibition held last year at the Oakland Museum. Left, a rare piece of Ceylon ebony, 7¼ in. across; right, osage orange bowl, 6 in. dia. Photos: Joel Schopplein.

tradition: he was not encumbered by utilitarian ideas about kitchenware. The shape of his turnings was startlingly new, as was their fantastic thinness. People admired them. But it is only from a distance of 40 years that we can see how they were the first of a new craft form.

Prestini was 25 when he took up woodturning. He'd apprenticed as a metalworker for two years, then studied mechanical engineering at Yale, and was teaching mathematics in Lake Forest, Ill. As he explained it in a recent interview, "My idea was to use making as a design resource, a different way to learn design. The usual way is to study the history of design, go to the library, have seminars. I tried to reverse things, to make the object first, then to draw it and think about it, then to make it again. This way a craft can teach you what information you have to communicate." In the late 1930s, the Bauhaus remnants under Laszlo Moholy-Nagy set up school in Chicago, then Mies van der Rohe came there to invent the glass-and-steel skyscraper. Prestini was in touch with both of them and became deeply immersed in the Bauhaus ideas about art and design, craft and industry. He says that in our technological era, his own humanity required him to develop the skill to turn wood as well as a machine could do it. But the shape of his bowls, which so impressed the Museum of Modern Art (p. 2), he shrugs off as easily derived from metal-spinning. "The important thing," Prestini says, "is not the product but the process. I'm not interested in turning a good-looking bowl, I'm interested in what does it take to turn that bowl? What do I have to learn to do?" And the main thing he learned, he says, "is that work is your best friend, it never fights back. I have reverence not for wood, but for work."

In other words, the real product was Prestini himself. In 1953 he began to work in metal sculpture because "there's so many things you can do with metal that you can't do with wood." His craft now part of him, Prestini-the-artist hires craftsmen who use advanced metalworking technology to build his sculptural conceptions.

During the late 1930s and throughout the 1940s, Prestini showed his bowls in museums and art galleries across the country, receiving considerable acclaim. Photographs of the bowls showed up in magazines, and the bowls themselves found places in many private and museum collections. People regarded their shape as an apt expression of streamlined modern times, and found it marvellous that they were turned from wood. They seemed as thin as china dishes, lighter than anybody had realized wood could be. Even so, serious woodturners who happen today upon an old bowl by Prestini see little that's remarkable. The woods are common birch, walnut, cherry and mahogany. A sharp eye finds sanding scratches. Many of them are variations of a single shape, a tautly convex curve from foot to rim, tilted more or less as the bowl widens. The thinness that seemed so magical when new is routine today, and nowhere near the limits of thin. The thing is, before Prestini perfected the techniques, nobody realized that such work could be done at all.

The professional—To the decorative bowl that Prestini discovered, Bob Stocksdale added the beauty of exotic woods from around the world (see pp. 41-45). And from that perfect silhouette, Stocksdale built both a family of shapes and also the skill to interpret a bowl's curve in terms of the wood itself. Where Prestini's bowls were in a way incidental to the process of becoming a designer, Stocksdale's purpose is more prosaic. Since the late 1940s, Stocksdale has been a professional turner of bowls and plates, his work at the lathe supporting his family. For many years and until quite recently, Stocksdale was probably the only professional turner of decorative bowls in America. Because of this, and because of the technical perfection of his work, Stocksdale has been an inspiration to dozens of aspiring young turners. He's been elected a Fellow of the American Crafts Council, and at 68 he's the grand old man of his field.

Although they live about a mile from each other in Berkeley, Calif., Stocksdale and Prestini are just barely acquainted. Stocksdale recalls seeking out Prestini about 25 years ago because he had admired the older man's work. Today Stocksdale says, "He's got only one shape, they're all the same shape."

The bowls shown above represent two of the shapes Stocksdale makes, in two of the hundreds of wood species he's turned. These bowls have become a technical standard in the craft—at exhibitions, you can see young turners studying Stocksdale's work and measuring themselves against it. Their wall thickness is between ⅛ in. and 3/16 in. There are no abrupt changes in thickness or in silhouette, inside or outside. There's no pimple or dimple at the center of the bowl, no torn end-grain anywhere on its surface, no sanding scratches, and no screw holes in the bottom nor any other trace of how the wood was held on the lathe (the secret weapon is a three-jaw chuck). The professional Stocksdale, working about 30 hours a week, can deliver about 30 such bowls a month.

The Exhibition

The innovators—Working from the basis that Prestini and Stocksdale created, a number of turners have invented or rediscovered aspects of the contemporary craft's vocabulary. Through the nine previous woodturning symposia and through the pages of this magazine, these innovators have freely given their discoveries and their techniques to anybody who wanted to know. Their generosity of spirit characterizes woodturners; it is one of the reasons for the rebirth of their craft.

Melvin Lindquist of Schenectady, N.Y., working with his son Mark, showed that you don't need to import logs from Africa to find exotic beauty. It's right there on the ground, inside partially rotted logs of New England maple, elm and birch. The Lindquists' early work was turned away from craft fairs, but their persistence—and the availability of new abrasives—unleashed spalted wood upon the world. Over the years they have perfected methods of turning this difficult material (see pp. 62-67).

Spalted wood can be turned only when the worker does not insist upon making functional kitchenware. It is a decorative material. But once you add spalted wood to the tradition of turning burl, and if you can accept mere existence as function enough, you can turn (and find beauty in) any bit of wood, no matter how worm-eaten, bug-infested, rotten or scabrous. These are either new ideas or newly popular ideas—I recently met an English master craftsman who was just shocked by the notion of turning rotten wood. After he'd spent a little time with a finishless, worm-eaten plate by Dale Nish (right), he came to agree that this new attitude could indeed uncover remarkable beauty.

Along with Rude Osolnik of Berea, Ky., Mel Lindquist was among the first to realize that a turned object doesn't require a pristine rim. Instead, its edge can reveal the original outside of the tree from which it came—a shape the turner does not create, he only selects and preserves. Then about the same time as David Ellsworth of Bucks County, Pa., Mel Lindquist rediscovered the 19th-century techniques of turning hollow, narrow-necked bottles through their neck openings (see Ellsworth's article on pp. 52-56).

Finally, Stephen Hogbin of Owen Sound, Ont., has shown how to escape from the lathe's circular nature, by cutting and reassembling turned elements into new forms (see pp. 68-72). Hogbin's work points always toward what else might be done, if the turner keeps eyes and mind open.

English walnut bowl *(10 in. dia.) by Melvin Lindquist combines the natural edge with the hollow bottle form. The turner cannot impose a preconceived shape on wood like this, but must see an appropriate shape within the log.*

Laminated mahogany and birch plywood bowl *(6 in. high) by Rude Osolnik. Osolnik, who with David Ellsworth chose the pieces shown in the Turned Objects Exhibition, is an exceptionally versatile craftsman. He is as comfortable with burls and gnarly roots as he is with the controlled materials and forms displayed here. But the various methods of laminating blanks for turning do not find much favor among the younger craftsmen, who, rather than build up layered blocks, would rather simply find and saw a suitable lump of tree.*

Photos, except where noted: Bobby Hanson

Worm infested ash bowl *(13½ in. dia.) by Dale Nish. A bowl from such awful wood can't be functional. This one was turned, and exists, for its own sake—for what it shows about wood, about worms, and about Nish.*

Walnut Bowl of Walnut *(10 in. high) by Stephen Hogbin. Photo: Staff.*

Box elder cluster burl *bowl (10 in. dia., 7½ in. high) hollow-turned by David Ellsworth. Like the blown shells of ostrich eggs, Ellsworth's hollow turnings can startle and amaze. It's difficult to imagine their lightness, or to believe that a piece of wood can be hollowed out through such a small opening. Nonetheless, this turning is completely hollow, with walls no more than 1/8 in. thick.*

The ideal bowl—Professional turner Bruce Mitchell of Inverness, Calif., won the "Best of Show" ribbon for his bay laurel flared bowl (7¾ in. dia., 5¼ in. high). Mitchell contrived to saw the blank, orient the axis of rotation and choose a shape that allowed the wood's fiddleback figure to squarely criss-cross its annual rings at every point on the bowl's surface, inside and out. The turning is correctly thin and skillfully flawless: no screw holes, no torn grain, no sanding swirls.

Juror Ellsworth said all of this still made it only another pretty bowl among many. But these qualities together with the precision of its taut silhouette made Mitchell's piece an archetypal symbol of the fond hopes of all the woodturners, and a perfect summary of the craft's recent evolution. Thus it won the prize, and most of the turners seemed to agree. But during the symposium debate several artisans declared Mitchell's silhouette a cliche, tritely modern, as dated as Prestini's shapes, nothing new and nothing inspiring here. Thus the bowl earned its prize again, by serving as the touchstone for a long and thoughtful discussion of the bowl-turning craft among the symposium participants: What good is a bowl such as this? Could you ever serve candy or pickles in it? Where would you keep it, besides on a pedestal? Is it purpose enough for a bowl to be a beautiful expression of nature's wonder, revealed by human skill? Does this make it art, or does art require newer and more profound insights? For myself, the bowl's beauty is its function, and I would give it a pedestal. But it is not the same as art. It is proud craft, admirably good craft, and that's enough for a bowl to be.

Best of show: Bay laurel flared bowl *(7¾ in. dia.) by Bruce Mitchell.*

Tulip poplar sphere bowl *(36 in. dia.) by Ed Moulthrop.*

Ed Moulthrop

Bowlus Fecundus *of Burma teak (9 in. dia.) by Marilyn Scott.*

Besides thinness—Turning a little bowl with coin-thin walls is a technical skill that separates the novice from the master. Thinness is now the aesthetic main line against which other types of bowl turning are judged. But there is more to this craft than thinness.

Ed Moulthrop of Atlanta, Ga., goes to the extreme of size. His tulip poplar spheroids range from 12 in. in diameter to 36 in., tree wide, always turned with the pith on the axis of rotation. They're usually the same squashed-sphere shape, fully hollow inside but with relatively thick walls. After rough-turning the green wood, Moulthrop stabilizes it with a long soak in polyethylene glycol, then completes the turning and sanding and applies an epoxy finish. Moulthroup is an architect turned professional turner who makes lots of these things. Whenever he puts together a gallery showing, it sells out. If you put your ear to the turning's opening, you can hear the sound the tools made when cutting into the whirling wood.

Marilyn Scott of Toronto, in this Burma teak turning she calls Bowlus Fecundus, confronts the notion that it has to be thin to be good. Although Bowlus is very thin at its rim, the walls get thicker as Bowlus gets deeper. Looking into it is like looking into a bell. You think you can lift Bowlus by grabbing its lip between thumb and forefinger, but you can't. You have to cradle it in two hands. Scott reminds us that mass and bulk are inherent characteristics of wood, whereas thinness is characteristic of porcelain or sheet metal. By embracing wood's mass, Scott has made a friendly fat thing to contain something special.

Carving goes well with turning, thereby generating another fresh universe of possibilities. Lottie Kwai Lin Wolff of Madison, Wis., carved the rim and interior of her thumbprint bowl, while Bill Hunter of El Portal, Calif., disc-sanded spiral flutes into his rosewood spheroid.

Del Stubbs of Chico, Calif., breaks away from the bowl by putting a tight-fitting lid on it. He's been making his living for four years by turning delicate little boxes; this one, in California walnut, is 3½ in. high. The lids of most turned boxes are smoothly sanded into the shape of the box itself, the join all but invisible. Stubbs avoids this easy way out, and accepts the difficult challenge of turning two related forms that fit harmoniously together. Part of the reason for his success is pure skill—Stubbs cuts surfaces and tiny beads that need only the finest sanding.

Bert Lustig of Berkeley Springs, W. Va., breaks the bowl upward, with a daringly deep vase form of black walnut, made special by its free edge. Lustig won a merit award.

William Patrick of Arlington, Vt., is among several professional turners who aren't stuck on the one-whole-piece-of-wood idea. Patrick sees the surface of a turned plate as a canvas for expressive drawing. He is exploring the color palette available in world hardwoods, and the shape-vocabulary made possible by the bandsaw; Patrick glues the picture together, then turns the plate. This example is zebrawood, mahogany, walnut, ebony and amaranth, 11 in. diameter.

Thumbprint bowl *of mahogany (7 in. dia.) by Lottie Kwai Lin Wolff.*

Spiral *(5 in. dia., 2 in high) by Bill Hunter.*

Bowl form *of black walnut (4½ in. dia., 8 in. high) by Bert Lustig.*

Covered jar of California walnut (5½ in. dia.) by Del Stubbs.

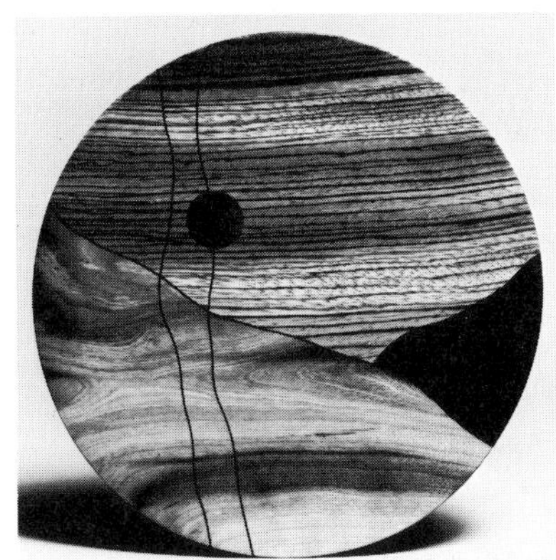

Plate *(11 in. dia.) by William Patrick.*

Vermilion bowl *(11 in. dia.) by Tom Eckert.*

Wall-mirror-table with clock *(28 in. wide, 20 in. high, 14 in. deep) by Robert Leung.*

Beyond bowls—As Scott's Bowlus enriches everyday life, Tom Eckert's vermilion bowl with bronze detailing and a skirt of orange feathers suggests a ceremony. The texture and color of the feathers against dark, purple-brown wood tempt you to touch, they make the bowl important, magic and mysterious: what relic lives here? May I peek?

Eckert, who teaches design and woodworking at the University of Arizona in Tempe, is no pilgrim at the shrine of the tree. He uses the lathe to explore forms and functions in other materials besides wood, in his own search for meaning. Turning technique and the wood's figure are secondary to Eckert's larger aims, so he doesn't mind a vertical glueline through the lid's knob. To me the glueline is a distraction that diminishes the mystery of the piece, making it seem ordinary again.

Robert Leung, a student in San Bernardino, Calif., assembled his wall-mirror-table from several koa-wood turnings. There's a clock at top center and two trapezoidal drawers under the shelf. Leung's turning suffers technically from end-grain tear-out, and his joinery is ratty, but the jurors still gave him an honorable mention. He deserves the encouragement, for Leung is not working in the well-known territory of bowl turning. He's plunging off into exploration of the lathe as a tool for making parts, a class of turning (and of furniture design) toward which Hogbin pointed but which remains unexplored. In this piece Leung has broken the sacred circle that came off the lathe, while restoring the circle by reflection in the mirror. The mirror glass peeping through the ring space in its frame makes ambiguous the boundary between glass and wood, between solid object and mirage. Leung is at the beginning of a journey,and it will be fascinating to see where he goes.

Translucent Goblet in Wood, *by Robert Street.*

The ghostly goblet—Bob Street of Aberdeen, Wash., an architect and amateur turner, turned this translucent goblet of Western ash, 7 in. high, 3 in. dia., 1/32 in. thick. It's a soft white ghost of a thing, weighing a mere 1.2 ounces. It has no finish, so it could never hold wine. All it can hold (along with your attention) is a yellow-orange glow when it's put next to the light.

Street's goblet came to be my favorite piece among the 100 Turned Objects in the exhibition. The wood he chose is the straightest imaginable, no hint of flashy grain. The form he chose is that of the common wine glass. The result transcends extravagantly figured wood and novel form by demonstrating that wood, the most rigid of materials, can achieve the delicate shapes of that most liquid material, blown glass. Thus it celebrates, in the humblest of materials, the limits of human dexterity. It was daunting to learn that Bob Street turned not just one goblet, but three of them, and then it was heartening to find out that the second and third took him much longer than the first. I'm glad I never saw more than one goblet at a time. □

Turning for Figure

Some design considerations when making bowls

by Wendell Smith

Recent articles about woodturning emphasize the techniques of turning. Discussions of grain and figure are generally found in books about wood or timber. The purpose of this article is to bring these two subjects together and to discuss some aesthetic aspects of bowl turning which depend upon grain and figure in wood. You can predict and control the figure that will appear in a completed piece by considering how your proposed bowl shape relates to the original orientation of the blank in the tree. The same blank can yield several types of figure, depending on how you orient it on the lathe and the radius of curvature you choose for the turning.

In this discussion, the words grain and figure will be used in the technical sense. Grain refers to the orientation of wood cells with respect to the axis of the tree. Authorities generally distinguish six types: straight, wavy, interlocked, irregular, spiral and diagonal. The first four are the most common. Figure is the surface appearance of the wood's anatomical features, including grain, which results from cutting or machining. In the same turning blank, tangential, radial and transverse surfaces display different figure, as shown on p. 10.

You can obtain many interesting effects using straight-grained wood by varying the orientation of the growth rings. Bowls can be turned with the growth rings "concave up" (the ring curvature running in the same direction as the bowl curvature) or with the growth rings "concave down" (with the ring curvature running opposite the bowl curvature). Some people find these relationships easier to visualize in terms of the bark side and the heart side of the wood. A bowl turned with the growth rings concave up opens toward what was the center of the tree, while one turned with the rings concave down opens toward what was the tree's bark.

Whether you turn with the rings concave up or concave down changes the appearance of a bowl or tray. I turned two cherry trays from blanks cut from adjacent segments of a single board, but I flipped one blank over before turning. A predictable pattern of concentric ellipses was produced on the tray blank oriented with the rings concave down (photo above left), while a hyperbolic pattern resulted when the rings were oriented concave up (photo below left). A striking example of hyperbolic figure is shown in the tulipwood bowl below. Turning concave up or concave down is a matter of personal preference, although I think that the former emphasizes flat-

---

Wendell Smith, who lives in Fairport, N.Y., is a research scientist and woodturner.

Photos: Wendell Smith

Cherry trays, left, from the same board show growth-ring figure from turning concave down (top) and concave up (bottom). Above, hyperbolic figure on tulipwood bowl by Brian Lee.

From *Fine Woodworking* magazine (November 1981) 31:75-77

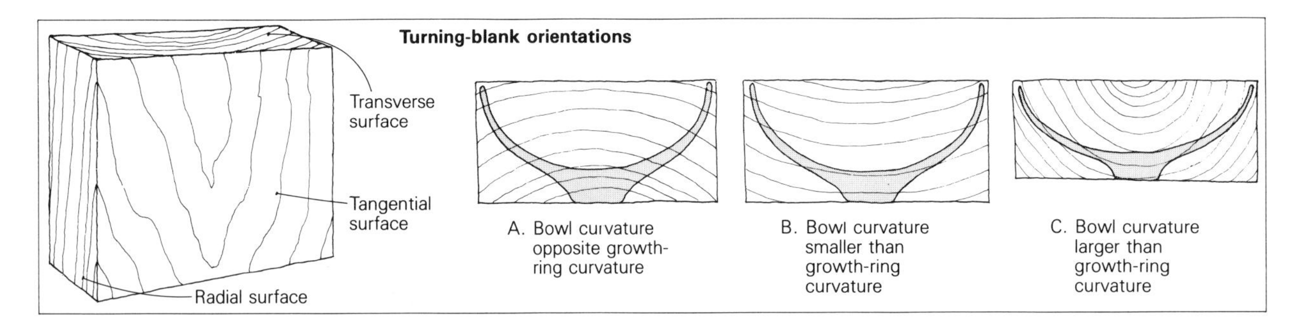

ness, while the latter accentuates roundness and depth.

These examples of growth-ring figure are really special cases of a more general rule. If the radius of curvature of a bowl is opposite to (A, above) or smaller than (B) the radius of curvature of the growth rings, a pattern of concentric circles or ellipses will result. If the radius of curvature of a bowl is greater than the radius of curvature of the growth rings (C), a hyperbolic and/or parabolic figure will result. Thus, under certain circumstances turning with the rings concave up can lead to either figure, as shown at right and below, on the butternut bowl. The base of the bowl has a greater radius of curvature than the rings, while the sides of the bowl have a smaller radius of curvature. Therefore, what begins as a hyperbolic pattern in the bowl bottom becomes a concentric pattern at the sides. The result is a set of rings that progresses across the bowl, a figure enhanced, in the case of butternut, by the characteristically scalloped annual rings of the species.

Many of the bowls shown here I purposely turned for symmetry. If symmetry is eliminated, the relationships discussed will still apply. Thus trial and error, and careful observation, can lead to turnings that fully reveal the beauty of the wood.

Two anatomical elements which are present to different degrees in all hardwoods are vessels and rays. Broad rays are characteristic of woods such as oak, beech, planewood and lacewood, regardless of the presence or absence of fancy grain. When rays are exposed on a radial surface, flake (or ray-fleck) figure results, which can be an interesting design element. Rays can be seen in radial and tangential section in the block of European planewood, left.

The planewood bowl, below, was turned from a quartersawn board. Because of this, the flake figure is revealed in the center of the bowl, but is lost toward the sides. At the lower left is a tangential (end on) view of the rays, and at the upper right, a transverse view. In fact, this photograph illustrates on a single surface of wood every conceivable ray figure that can be exposed in European planewood.

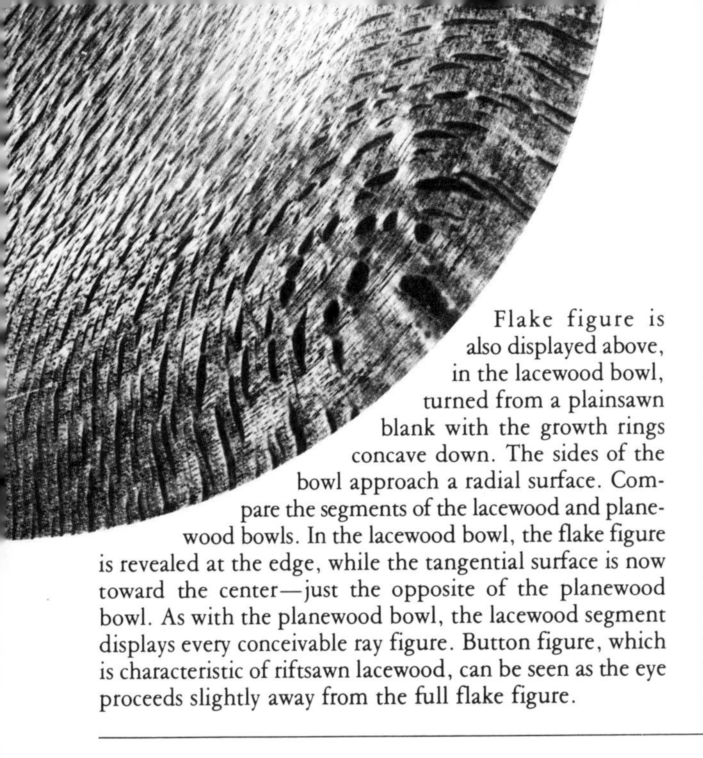

Flake figure is also displayed above, in the lacewood bowl, turned from a plainsawn blank with the growth rings concave down. The sides of the bowl approach a radial surface. Compare the segments of the lacewood and planewood bowls. In the lacewood bowl, the flake figure is revealed at the edge, while the tangential surface is now toward the center—just the opposite of the planewood bowl. As with the planewood bowl, the lacewood segment displays every conceivable ray figure. Button figure, which is characteristic of riftsawn lacewood, can be seen as the eye proceeds slightly away from the full flake figure.

In transverse sections of certain species of hardwoods, such as elm (sometimes hard to find because of Dutch elm disease) and osage orange, the latewood pores appear as wavy, concentric bands. A tangential surface which cuts across the tops of the waves will lead to a figure resembling bird feathers, which is called partridge-breast figure. This figure is illustrated in the elm tray, above. I turned this tray with the rings concave up because the surface of the tray would be very close to tangential. In a perfectly tangential cut the figure is rather diffuse.

Orientation is also important with woods having a fancy grain, such as wavy or interlocked. Wavy grain leads to a figure known as curl, which is best displayed on a radial surface. The curly maple bowl, left, was turned with the rings concave down. Consequently, the figure is slightly more pronounced at the sides of the bowl (radial cut) than at the base (tangential cut).

The same considerations apply to turning woods with an interlocked grain as apply to those with wavy grain. Interlocked grain, which is typical of many tropical woods, occurs when fibers of successive growth layers spiral in opposite directions. This results in a figure on quartersawn surfaces known as ribbon, or stripe, of which striped African mahogany is probably the classic example. To display this figure in a plainsawn board it is best to turn with the rings concave down. Alternatively, the combination of a quartersawn board with maximum radius of bowl curvature would also lead to maximum stripe figure. The combination of interlocked with wavy grain can lead to several other figures, many of which are better displayed on a radial surface.

Irregular grain is the fourth type considered here. One example, caused by localized swirls in the grain, is bird's-eye. Because bird's eyes develop radially within the tree, a tangential cut best reveals the ends of the eyes. Thus, turning concave up may be preferred, as in the bird's-eye maple bowl at left (turned by Al Stirt). On the other hand, bird's eyes can also be attractive in cross section, which calls for cuts that expose a radial surface.

The most frequently encountered irregular grain occurs in the vicinity of crotches. Boards which at first sight might be rejected because of unwanted knots can be used to turn beautiful objects. A blank removed from a board by sawing close to but not including a knot will usually lead to a figure with chatoyant swirls. The ultimate in crotch figure is generally taken to be a full-feather crotch, shown at left. To obtain this figure, however, bowl design really must begin when the log is cut. For more on cutting wood for figure, see "Harvesting Green Wood" by Dale Nish, pp. 14-17. □

The Bowlmaker

The turner's art in Ethiopia

by Chuck and Nancy Boothby

The Bowlmaker spent several weeks each year in Mendi, a town in western Ethiopia near the Sudan, where we taught English during 1968 and 1969. Mendi, three days drive from the capital, was the last town in this direction with an ample supply of cordia—most of these trees closer to the capital had been cut into doors and window shutters. This was his preferred wood—easy to carve, with a beautiful, dark, rich, wavy grain.

The Bowlmaker would come to our town in the dry season, buy a large tree, cut it into logs and turn the wood in the shade of a nearby front porch. The green wood turned easily. When a piece was completed, the Bowlmaker would finish it with animal fat mixed, intentionally or accidentally, with red clay. When we first saw the Bowlmaker, we were fascinated with the simple beauty of his work. Later on, while taking the photographs that appear here, we were captivated by how his body, his equipment and the natural environment all fit together into an aesthetic unity—quite unlike that of an erect, modern electrical lathe wobbling on an irregular concrete floor inside a stark, dusty building.

In his work, the Bowlmaker used ax/adzes, turning tools and drills—all made by a local blacksmith using primitive leather bellows with cow-horn nozzles. He had three ax/adzes: a large one for chopping and roughing out, a smaller one for fine shaping and one with a curved blade for rough-hollowing the bowls. All three blades had hollow, tapered throats for the pointed end of a crooked handle. While this is probably the easiest way to attach a blade to a wooden handle, it is also the most practical—the blade is easily removed to change ax into adze as the job requires. Hitting the curved part of the handle on its side loosens the blade; hitting the handle with the blade up tightens it.

His turning tools were long metal shafts embedded in handles made from sticks that were 1 ft. long and 2 in. wide. Unlike our precise, machine-made tools, the blades were shaped to the curves useful in his work. The drill, which was worked by rubbing it rapidly between the palms of the hands, was a ⅝-in. diameter stick, its bit a piece of metal ⅛ in. in diameter with the point flattened on the sides to form the cutting edge. The Bowlmaker used two sizes, one about ⅛ in. and the other about ¼ in.

Many of the Bowlmaker's designs were influenced by foreigners, especially missionaries. For example, his magnificent candlesticks had a definite cross shape. Similarly, his lidded bowl/table (for family sharing of a spicy stew served on large sourdough pancakes) collapsed so ingeniously that it seemed to have been designed specifically for the suitcases of foreign visitors, though it was an integral part of local custom. The Bowlmaker first approached us with his usual product line, but when we described a set of salad bowls instead, he eagerly took on the challenge. The idea of a set of bowls all exactly the same size, however, was a strange concept to his way of thinking. He turned bowls in the same style, but their dimensions were determined by the width of the portion of tree limb used for the stock, as well as by the Bowlmaker's varying aesthetic energy. He used no standard measuring tools, but gracefully measured with fingers or forearm, then remembered the results. More precise measurements were made by breaking the nearest straight twig to the proper length, and pocketing it for later reference. □

Chuck and Nancy Boothby teach at Moorestown Friends School, Moorestown, N.J. They built their own summer home in Sedgwick, Maine.

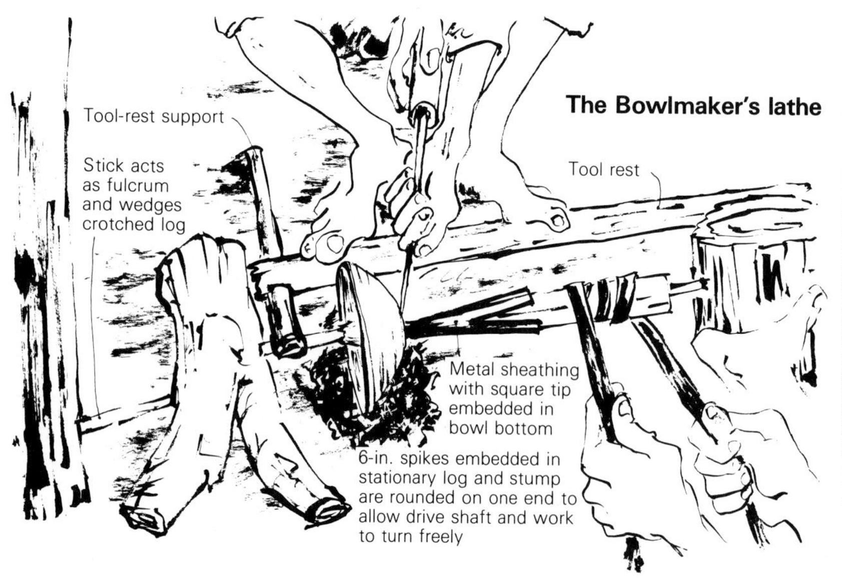

Photos: Nancy Boothby

The Bowlmaker's specialty—a collapsible lidded bowl/table for the family meal.

From *Fine Woodworking* magazine (March 1980) 21:54-55

The Bowlmaker splits a large green log in half lengthwise with an ax/adze, then cuts it to length. Rounding the ends completes external shaping.

The next step is rough-shaping the inside of the bowl with a concave-bladed ax/adze. The knob in the center is left intact for attaching to the headstock of the lathe.

Next he drills a center hole in the bottom of the bowl—the drill is turned by rolling between the hands, while the Bowlmaker steadies the bowl with his feet.

The headstock, a stick with a cone-shaped end wrapped in metal and square at the tip, is banged securely into the center hole with a mallet. Then he marks the tailstock center with a smaller drill.

The lathe is powered by a strong assistant pulling forward and back on a long leather strap wrapped around the headstock. For symmetrical cutting, the strap must be pulled rhythmically.

Left, the Bowlmaker turns the inside of the bowl—little sanding is necessary because of the smoothness of the cuts. To sand, the Bowlmaker uses local abrasive leaves—fig leaves, according to a botanist in Addis Ababa. When the turning is complete, above, the bowl is taken off the lathe, with the center knob still intact. Knobs on both top and bottom will be removed with a small ax/adze.

Rough areas are smoothed with a few strokes of the turning tool, and the bowl is done.

Harvesting Green Wood

Patience and perseverance pay off

by Dale Nish

I have been working with green wood for several years. I have found the satisfaction of finding the wood almost as fulfilling as actually working it. In some cases, as with life, the expectations and anticipation are the most rewarding part. Once while deer hunting, I found a large, soft maple burl growing at the base of a small tree. The burl encircled the tree, rising to a height of perhaps 30 in. with a diameter of over 48 in. Over several years I envisioned the turnings which could be obtained from a burl of such beauty and size. At last I received permission to cut it off, and gathered together the necessary tools and equipment, a major undertaking, as the burl was far off the main road. After arriving at the tree and inspecting it carefully, an enthusiastic friend and I began to saw—the chain was sharp and the chips flew. Suddenly the bar fell into the burl. It was completely hollow. The only sound wood was a 2-in. to 3-in. shell. The burl was of no value at all for turning. Maybe next time.

Photos: Max Wilson

These bowls were turned from green boxelder burls found in Nish's neighbor's firewood pile. They measure 7 in. to 15 in. in diameter.

Seasoned and finished bowls turned from green silver maple, apricot, ash, black walnut, lignum vitae, macassar ebony, honeylocust and spalted maple.

The sources of wood for turning or carving are limited only by your patience and perseverance. The best wood cannot be purchased from a lumberyard or hardwood dealer, and even if the desired species is available, you will still be limited by the sizes offered. Most of us live where species growing locally far exceed species available commercially, but if you want to work local wood, you must cut your own.

Wood is everywhere. Robert L. Butler, in his book *Wood for Wood Carvers and Craftsmen* (A.S. Barnes Co., Inc., Cranbury, N.J. 08512), has a chapter aptly titled "Wood Is Where You Find It." In Utah, which is not noted for its forests, I have harvested locally grown oak, ash, maple, black locust, honeylocust, mulberry, English walnut and black walnut, American and Siberian elm, ailanthus, catalpa, cottonwood, poplar, boxelder, aspen, chestnut, sycamore, apple, pear, cherry, plum, peach, apricot and more. Wood is everywhere. I have found it in firewood stacks, trees bulldozed to clear building sites, limbs left from logging operations, windfalls after a storm and orchards being uprooted. Other good sources of turning and carving wood are tree-removal companies, city shade-tree departments, local sawmills, landfill or dump areas and friends and neighbors who know you are a wood nut and inform you when they see trees being cleared. Local sawmills frequently have short or crooked logs which have been discarded as uneconomical for processing into lumber. These logs are either inexpensive or free. Show appreciation with a gift of a turning or two and see how your supply increases.

A minimum of equipment is required for cutting your own wood. A chain saw is a necessity, and you will also need a maul, splitting wedges and a peavey (if the trees are large). A pickup or trailer is handy, but you will be surprised how much wood you can haul in the trunk of your car.

The chain saw must be sharp and in good condition. Cutting parallel to the trunk of the tree is different from cutting cross grain, as the chain cannot cut efficiently across the end surface of the log. I use a chain saw with a 16-in. bar for most of my cutting, but I have a saw with a 30-in. bar for larger pieces. Here in Utah it is uncommon for a tree to be more than 30 in. in diameter.

The chain must be sharpened according to manufacturer's specifications—teeth even in length and equally sharp, or the saw will lead toward the sharp side of the bar. If the saw has an automatic oiler, check to be sure it works. The oil reservoir must usually be filled each time you fill the saw with gas. Manual oilers must be used frequently, as improper oiling or insufficient oil will raise havoc with the bar and chain. Always

From *Fine Woodworking* magazine (May 1979) 16:48-51

use ear protection, since many saws can cause permanent ear damage after a short time of continuous use.

Cutting the wood — The first step in working a bolt (log section) is cutting off its ends to remove end checks. If you are concerned about nails or dirt in the bark or wood, remove the bark with an ax. This isn't always necessary, and I usually leave the bark on the log until I am ready to work it. The bark helps keep the log from drying out and checking, but it also encourages grubs and beetles, which may ruin the log or at least destroy the sapwood. If you remove the bark, cover the log with plastic to prevent drying and checking.

After the end checks are removed, measure the useful diameter of the log and cut the bolts in lengths equal to the useful diameter or in multiples of it. Don't cut short lengths unless you're ready to work them. Short lengths quickly check and the bolt may be ruined.

After a bolt is cut, I stand it on end on wood blocks or slabs. Be sure the bolt is in a stable position for sawing. The bolt may be laid out different ways, depending on defects, pith (small growth center of the tree) position and end use. Large checks may sometimes be the places to make the cuts. One thing is constant—the pith must be removed from the flitches (blocks of wood cut from the bolt). Several options for layout and cutting are shown in the diagram at right.

Once the cuts have been outlined on the end of the bolt, make cuts parallel to the sides of the flitches. These cuts should remove most of the bark, but don't cut too deeply into the sapwood. The exception here is fruitwoods. Their sapwood is almost impossible to season without checking and should be removed.

The next step is the primary cut, which will usually halve the bolt. For the remaining cuts remove the pith. Any remaining bark should be removed with the ax.

Try to cut with the sawbar making an angle of about 30° to 45° to the end of the bolt. Cutting parallel to the end of the log is inefficient because you are cutting end grain, and cutting parallel to the length of the bolt produces long shavings that can not clear the chain, causing it to bind and overheat. Short bolts can be cut standing on end. Long bolts must be laid down for sawing. In either case, the bolt must be in a secure position and raised sufficiently to allow the chain cutting room without contacting dirt or rocks.

Both the bowl and the block contain the line resulting from grafting English walnut to Claro walnut root stock. The block was obtained from a tree grown in northern California.

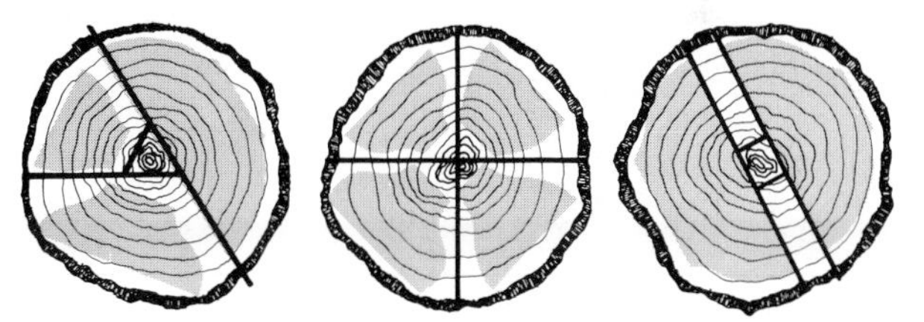

Lay the bolt out according to how many bowls will fit, pith and any defects.

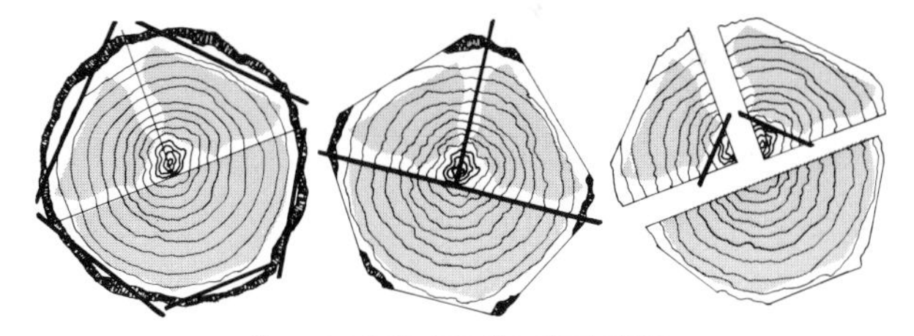

To cut a bolt, trim the sides of the flitches (left), make the primary cuts (center), then slice off the pith (right).

Cutting for figure — Most of the pretty figured wood in a tree will occur in the area below the major fork (crotch figure), in the stump area (stump figure) or in the occasional burl on the trunk or around the base of the tree. Crotch figure, the most beautiful, is seldom found commercially because it is usually trimmed off at the mill, or is so thin or short as to be of little value. Some of the finest crotch figures I have found came from local cottonwood trees on their way to the dump. Cottonwood trees, at least in Utah, have little commercial value and are seldom used even for firewood. Other species with beautiful crotch wood are honeylocust, black walnut, elm, ash, catalpa, aspen, cherry and apricot.

If I have a complete tree, I try to work the major fork first. The first two cuts, after the small limbs and branches have been removed, are about 24 in. above the fork of the tree, severing the two primary limbs from the main trunk of the tree. If weight is on the limbs, it's good practice to cut up from the bottom side to a point about halfway through, then cut down into the limb at a point 2 in. or 3 in. away from the first cut, toward the top side of the tree. As the downward cut progresses the cut should open up. Be careful, as the limb may break down to the first cut, and may roll or twist toward you. The next cut is made 24 in. to 36 in. below the fork of the tree. Before the cut is made, be sure to support the trunk with wooden wedges or blocks to prevent the trunk from settling and binding the saw blade.

After the fork has been cut from the tree, examine the ends for decay, splitting, insect activity or other deterioration. Use a lumber crayon, and mark the pith on the ends of the fork. If the fork is sound, transfer the crayon marks to the sides of the fork. Use a straightedge to connect the marks from top to bottom. These lines will be the cut lines when the fork is split.

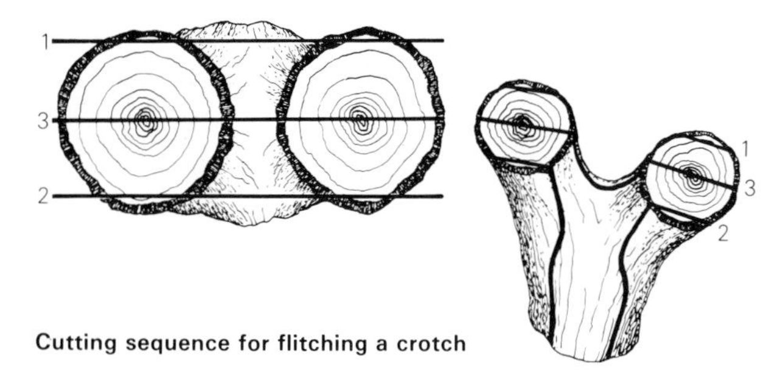

Cutting sequence for flitching a crotch

Using the center line as a guide, saw a slab from both sides of the fork, trying to keep the cut parallel to the center line. The slab cuts should remove part of the sapwood and bark, but should not be so deep that they remove wood that could be used during turning. Watch for nails or other metal, and pull or chop it out of the wood. Expect to hit a few nails if the trees come from yards or fence lines. Sharpening a chain is a small price to pay for a quality piece of wood, and in my experience, a nail or piece of wire does much less damage to the chain than does a small rock or pocket of sandy dirt.

After the slab cuts are complete, saw down the center line. Do not try to saw straight across, parallel to the ends of the forks. Rather, angle the cut 30° to 45° to the end-grain surface. This allows the chain to cut more efficiently. Too little an angle and the wood cut out by the chain will resemble sawdust; too much of an angle will cause the chain to produce long slivers or shavings, which will clog the chain drive. Experience will show you the best cutting angle. If the fork is large you may have to cut from both sides, at least until you reach the main trunk below the fork. Start the cuts carefully and be sure they line up and will meet at the junction of the fork. A cut running off to one side will require extra work, and could ruin a piece of wood of exceptional beauty and value. A perfect cut would be right down the pith of the tree, leaving part of the pith in each piece, resulting in minimum waste and two true, flat slabs. It is always exciting to make this cut and watch the fork separate. A feeling of wonderment comes over me, and I can hardly wait to see what is revealed. It is better than Christmas, because it happens every time a piece of wood is worked. One cannot do this and not see the Master's hand in this beauty of nature.

Wedges are used to split a cherry stump containing pockets of dirt and rocks. Be sure to split through the pith.

Some trees, such as black walnut, will have crotch figure anyplace a branch attaches to the trunk or larger branch. Other trees may not have true crotch figure, but you can always find beautiful figure in that area. Crotch figure of small trees is thin, often only an inch or two deep. To preserve it, shallow trays or plates must be turned, with the figure at the bottom so one can turn down to it and reveal it. Otherwise, the turner will go through the figured area into the plain wood, leaving the figure visible only at the edges.

Stumpwood — Stumpwood is that portion of the tree which starts to flare at the base of the trunk and continues into the ground. A tree is usually cut off at a point 1 ft. to 2 ft. above the ground, leaving the stump intact. One can often saw the stump off close to the ground and collect a fine piece of wood. At other times the stump may be dug out completely. The roots are cut off and the remaining stumpwood cut up. For the woodturner the stump offers many opportunities, including a lot of hard work. Color and patterns are beautifully innumerable. Stumps contain both sapwood and heartwood, just as in the trunks of trees, but they will not be so evenly separated and will flow into each other. Colors are often more dramatic, with streaks of black and dark browns coming to life, producing marble-like patterns. Because the stump is in continuous contact with the moist ground, it is subjected to mineral stains, stains from decaying surface matter and other colorants. The irregular grain found in the stump may show as quilted figure, fiddleback, ribbon or swirl—the only thing certain about a stump is uncertainty. You can expect figure equal to or better than that found in other parts of the tree, with the possible exception of the crotch figure. Because stumpwood figure goes all the way through the blocks, stumpwood usually can be cut into various sizes and turned with little regard to grain direction. This wood is excellent for deep bowls of simple design.

Burls — Burls are rare, and usually available only as veneers, but if the woodturner is alert, burls can be found. Burl wood has most interesting designs, in many colors and textures. These wart-like growths are usually found around the base of the tree, but may appear anywhere on the main trunk. The burl consists of a mass of dormant buds, sometimes called eyes. Therefore, there is no alignment of wood fibers, and the burl is gnarled and misshapen. This in turn produces figure of unpredictable color and pattern. Because burls have no grain direction, they are quite stable when turned green, and in most cases are easier to work than wood from other parts of the tree. Make sure to use sound burls, without bark pockets or decay. Cut them into their most useful sizes, disregarding grain direction. Often only small pieces can be turned, because of defects, but if the defects are small, I often turn the piece and leave the defect in the surface of the turning.

Treating green flitches — Trees are designed to carry sap, and so long as a tree is alive, its cells are full of water. When a tree is cut down, it begins to lose moisture. This process is called seasoning or drying. As the wood loses water from the cells it becomes lighter, harder and stronger, and it also shrinks in size. Seasoning will continue until a balance is reached between the water in the wood and the moisture in the air around it. This balance is called equilibrium moisture content (EMC). Because the EMC will vary with the humidity

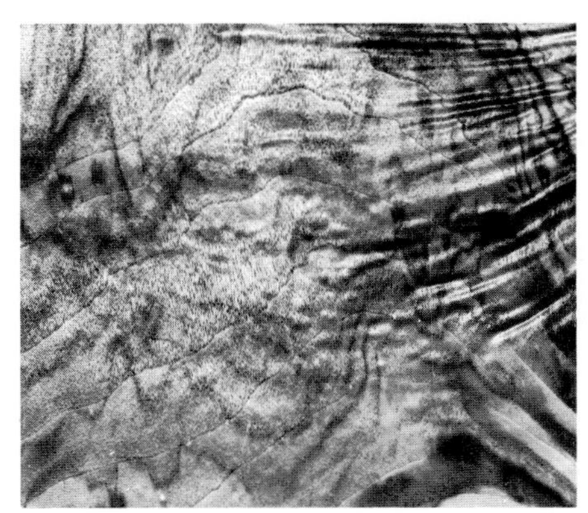

Marble-like figure of walnut stump can be turned with little regard to grain direction.

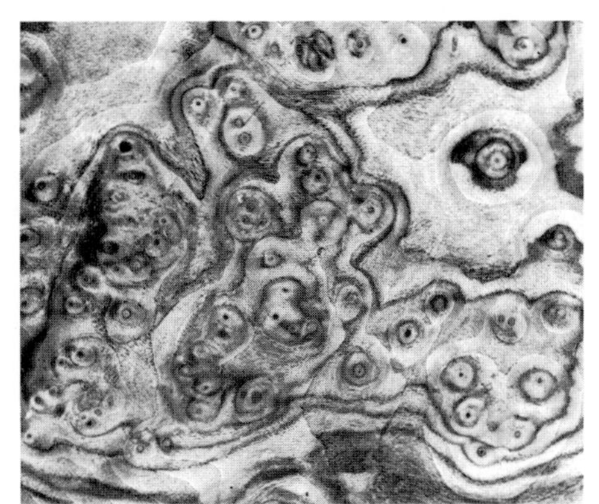

Walnut burl. The eyes are formed by the dormant buds that compose the burl.

Shallow trays were turned from ash with crotch figure.

in the surrounding air, final seasoning should be done in areas similar to where the wood will be used. In most cases, this will be in heated rooms.

Green flitches should be end-coated immediately after cutting, as checks will quickly appear unless the ends are sealed against moisture loss. If the flitches are to be seasoned before turning or held for a month or two before turning, I prefer a commercial end-coating made by Mobil Oil Co., called Mobilcer-M. Coat the ends and about 2 in. in from each end, and also areas of high figure, knots and (with some species) sapwood. Other end-coatings such as hot paraffin wax, asphalt, thick oil-based paints, vaseline or white glue may also be used. If a flitch has been cut and left for a few days without treatment and the ends are checked, make a fresh cut to remove the checks before end-coating. For temporary storage of flitches or green turning blocks, I often use plastic bags.

Flitches that are to be seasoned for later use should be treated much differently. I cut them as long as possible, because there will be less trimming waste when they are finally used. I then weigh each piece and write its weight and the date on the flat side of the flitch. The weight will be recorded periodically, and when it stabilizes, the moisture content of the wood will be in equilibrium with that of the atmosphere.

After weighing and end-coating the flitches, I stack them in an unheated shed. Sticker the flitches if flat and of uniform thickness, or stack them if they're of random size and thickness, to allow for good air circulation. Then cover them tightly with a plastic sheet. In humid areas, you could probably leave them uncovered.

If you weigh a block of green wood on a daily basis, you will note that most of the weight loss occurs in the first few weeks. This is also the time when checking is most liable to occur. Covering the green wood with plastic gives it a chance to season slowly, without checking. If the wood has a lot of figure, or is very valuable, I sometimes leave the plastic on for several months and then open the bottom of the cover to allow the direct outside air in contact with the wood. Over a period of a month or two the cover is opened more and more until it is completely removed.

When the weight has been stable for several weeks, the EMC has been reached. In most areas, this will be between 12% and 15% moisture content. At this point, the wood should be brought into a heated storage area and allowed to season to between 6% and 8% moisture content. The time necessary for this depends on species, temperature and thickness. Periodic weighing will indicate when the moisture in the wood has reached the EMC of the heated room, or a good moisture meter can be used to check the wood.

Wood seasons at various rates, but you can expect at least a year per inch up to 8/4 stock, and three to four years for 12/4 or 16/4. Often, thick stock takes five years or more to season, and even then is not suitable for finished turnings. Green turning is the best solution for working thick wood.

If it is necessary to use wood that has been seasoned in an unheated area, rough-turn the bowl to shape, leaving a wall thickness of ½ in. to ¾ in. Remove the bowl from the lathe and allow it to sit on the bench for a week or two. It will probably warp a little, but it should not check. If you see checks appearing, put the bowl in a plastic bag for a few days to allow the moisture content to stabilize. Then, remove the bowl from the plastic bag and allow it to continue seasoning. Unless the wood has a high moisture content, one treatment in a bag is usually enough.

If during the turning process the bowl seems really damp, and you can feel the moisture in the wood, complete the rough turning and treat the bowl as green wood.

After you green-turn — Green wood should be cut to rough shape, mounted and turned in a manner similar to turning seasoned wood—except it is much more fun. After the bowl is green-turned to a uniform wall thickness of ¾ in. to 1 in. for bowls less than 8 in. in diameter, and 1 in. to 1¼ in. for larger bowls, I coat the surfaces, inside and out, to control checking during seasoning. Before coating the bowl I often weigh it and write the date on the bowl.

For bowls with no problem areas, a heavy coating of paste wax is usually sufficient. Coat the end grain carefully, forcing the wax well down into the fibers. If a bowl has high figure, knots or sapwood, coat these areas with Mobilcer-M, let the coating dry until it has a clear appearance, then coat the remainder of the bowl with paste wax. Place the coated bowls on the floor or on low shelves in an unheated area with little air movement. After about a month, move them to a moderately heated room. Bowls coated with paste wax will season and reach equilibrium moisture content in about three months. If I am in no hurry, I often dip bowls in Mobilcer-M and let them drain dry. These bowls will take six to twelve months to be ready for finish-turning. □

Dale Nish teaches industrial education at Brigham Young University, Provo, Utah. He is the author of Creative Woodturning, *published by the Brigham Young University Press.*

Vermont Turning School

Russ Zimmerman's three principles for clean cuts

by Dick Burrows

Like many woodworkers, I learned turning with a book in one hand and a gouge in the other. Whenever I read about a slick technique, I'd imitate it, seeking those satisfying cascades of shavings and mirror-smooth finishes. Most often I'd catch the tip of the tool and brutally slash the wood. One night, while holding an ice pack to my jaw, pondering how a shattered bowl could hit so hard, I decided to temper my quest for world-class cutting technique. When things got risky, I'd put the skew and gouge away, grind a burr on my ½-in. roundnose and scrape.

My turning gradually improved anyway, largely because I repeatedly practiced what Dale Nish and Peter Child had written. No matter how good the finished pieces looked, though, I didn't like the hit-and-miss combination of cutting and scraping I used to hack them out, or the pieces I broke trying. Something was wrong, but there weren't any good turning teachers nearby to help. Wouldn't it be great, I thought, to attend one of those intensive seminars, like the ones Child conducted at his home in England, and learn a better way?

Unfortunately, Child retired from teaching before I could visit him, but I recently spent a couple of days with one of his students, Russ Zimmerman, who runs a turning workshop in Westminster, Vt. Zimmerman modeled his school after Child's, but has modified many of the methods he saw Child use in the mid-70s, because of his subsequent experience and his contacts with other turners. As Child did, Zimmerman limits each class to two students, who usually move in with his family for 2½ days and spend most of their waking hours turning. My fellow student, Nils Agrell, a New York City stock investor, and I each had our own Myford lathe to work on in Zimmerman's compact and efficient basement shop.

One of the first things I noticed about Zimmerman's turning is that it's much more relaxed than my mish-mash of techniques. Where I would strangle the tool, jam it in and hang on, Zimmerman stresses control more than brute force. Steadying the tool against his leg or hip, he uses his whole body to move the tool, adjusting the cutting edge with light hand and finger pressure, taking full advantage of the tool's bevel and cutting edge.

Zimmerman bases his turning on three general principles—the cutting edge should be about 45° to the direction in which the work is rotating, the tool's bevel should be rubbing on the wood, and finishing cuts should be made across, not against, the wood fibers. These principles are presented as guidelines for developing a feel for tools and an understanding of what they are doing. Zimmerman urges students to build on this understanding, and to ask themselves what he thinks is the most important question: "Does it feel right?" His only iron-clad *must* is sharpening. The grinder, shown on p. 20, is one of the most used tools in the shop.

Zimmerman demonstrated his principles by hollowing a small walnut bowl using a long-handled, ¼-in. deep-fluted gouge. With the bevel riding lightly against the wood, but not cutting, he moved the tool forward and shifted the tool handle slightly until the gouge began to cut, then moved his body to continue the cut across the wood's surface. He steadied the handle with his side and kept the bevel rubbing while manipulating the angle of the cutting edge. Since I manipulated turning tools with just my arms and hands, I initially thought Zimmerman's style of work was more like dancing than turning, but I soon found that by bracing the 15-in. handle against my leg or hip and moving my body and tool together, the tool's cutting edge was much more stable and easier to control.

After stressing the importance of body movement and bevel contact, Zimmerman urges his students to play with the techniques. He convinced me that I needn't worry about the edge digging in, as long as the bevel was in contact with the wood. Being overly cautious, I practiced for a while with the lathe motor off, turning the wood into the tool by hand. Even with the lathe on, I was surprised at how easy it was to adjust the tool angle and cutting edge, as long as I kept the bevel riding on the wood.

The recommended 45° angle of the cutting edge to the work makes sense if you consider that when the edge is at 0°, just about parallel to the direction of rotation, no cutting takes place. Adjusting the handle slightly to a 15° edge angle produces a light cut and a fine finish, because the bevel is able to polish the surface behind the cut. As the angle increases to 90°, as shown on the facing page, more of the edge contacts the wood, and the cut becomes heavier and more likely to tear the endgrain. The 45° angle compromise produces a good surface and a reasonably fast cut without requiring excessive force. Once the position of the cutting edge is set, Zimmerman describes the angle of the gouge's flute in terms of how a clock face would look with a gouge cross section superimposed in the center of the dial. A straight-up flute would be 12 o'clock; one parallel to the tool rest would be 3 or 9 o'clock. As the flute orientation changes, note how a different part of the edge begins cutting.

The hardest thing about manipulating the bevel is developing a light enough touch to skim the bevel across the wood, instead of rubbing it so hard that it burnishes the wood, heating and dulling the tool. It takes time to develop a feel for moving the handle and cutting edge simultaneously at the same rate. If you shift the handle without moving the tool forward to keep the

From *Fine Woodworking* magazine (January 1986) 56:40-42

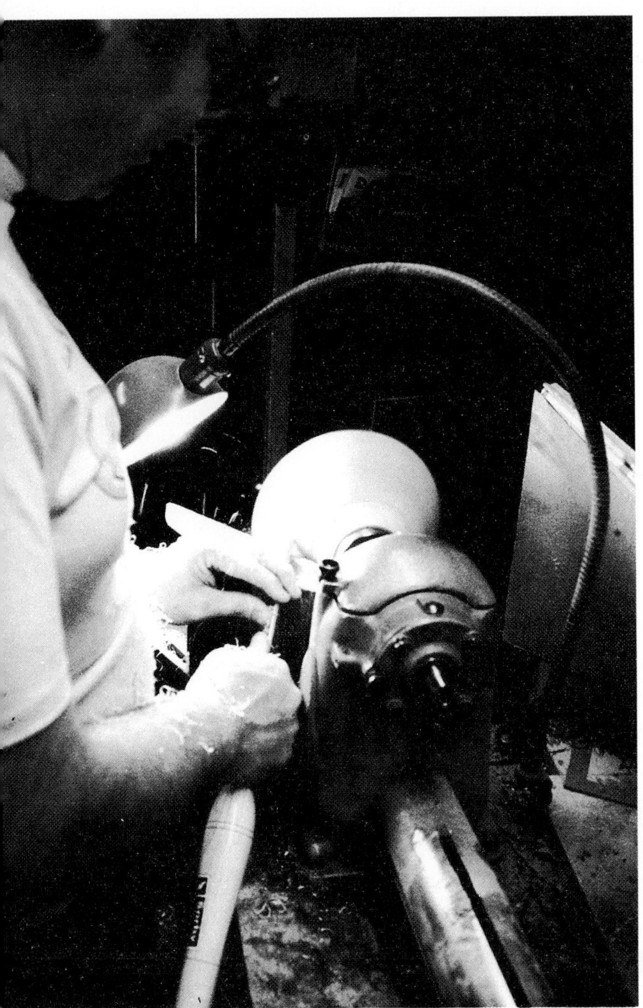

Fig. 1: Flute angle

Flute angle is described with a clock face. 12 o'clock is good for roughing out; 3 o'clock better for starting a hollow cut.

12
9
3
6
12
9
3
6

Fig. 2: Chisel angle

Direction of rotation

Chisel angle regulates size and smoothness of cut.

15° Fine cut

90° Roughing out cut

45°

Turning student Nils Agrell, above, braces the tool handle against his hip to steady the cutting edge, while he manipulates the bevel and flute angles with his hands held high on the tool. To hollow a bowl on the lathe's outboard rest, right, adjust the flute to the 9 to 10 o'clock range and skim the tool bevel along the cut surface as you arch the chisel deeper into the blank. A fingernail gouge rides its bevel up the outside of a bowl, far right, smoothly cutting across the wood fibers. Keeping the bevel against the wood at all times prevents the cutting edge from digging into the stock.

bevel skimming the wood, you will either lift the edge off the wood, stopping the cut, or the edge will dig in and the cut will be rough and difficult to control. A rough cut creates a bump, on which the bevel bumps again next time around, cutting another bump, and the bumps will reproduce quickly. Stop, go back to the last smooth area, set the bevel and recut.

Agrell and I had our choice of doing spindle or bowl turning, or both, and each of us elected to spend most of our time on bowls. We began by making a small walnut bowl. We each worked a 4-in.-thick bandsawn blank screwed to a faceplate on the outboard spindle, first turning the outside shape, then flattening the bottom with a ½-in. deep-fluted gouge. Flattening the bottom was a good chance to practice body movements, since the straightness of the bottom depends on the handle moving at the same rate as the cutting edge. Begin the cut with the bevel parallel to the bottom of the bowl and the flute facing 10:30. Then, brace the end of the handle on your leg, set the bevel, and use your leg to push the tool across to the center.

During each cut, Zimmerman makes his students watch the emerging shape on the bowl in the area opposite the actual cut. If you concentrate on the cutting edge, you'll instinctively try to keep the tool cutting and never notice the character or flow of the shape you are cutting. I found it hard not to focus on the tool, so Zimmerman repeatedly put his hand in front of my bowl, blocking my vision. It's uncomfortable not being able to see the business end of the tool, but he was right—the shape was more flowing and elegant when I didn't just stare at the tool.

Before removing the bowl from the faceplate, we held a second faceplate against the rotating wood, centered it by eye, and held a pencil to the wood at the faceplate's rim. The drawn circle is the guide for remounting the blank. It's a surprisingly accurate method. Instead of screwing the faceplate to the now flat bottom of the blank, we attached it with Permacel double-faced cloth tape. Cover the faceplate with tape, peel off the

Grinding turning tools

Zimmerman insists that a key part of turning is continually sharp tools. He uses a Sears' bench grinder, right, fitted with custom-made tool rests and guides. The two tool rests are 3-in. by 4-in. by ¾-in. pieces of plywood bolted to the grinder's original metal rests. The right-side rest, which is used for skews and parting tools, is angled to produce a 30° to 35° bevel on the tool. Scrapers are ground on the left-hand rest, which Zimmerman sets to produce an 80° bevel.

The third bench-grinder modification is an adjustable tool-handle support for grinding square-edge and fingernail gouges. The 5/16-in. by 1¼-in. arm slides in a small box mounted to the table under the wheel. A thumbscrew locks the arm in position. The tricky part is setting the height of the rest to fit your tools.

Zimmerman uses a sliding tool support to steady a long-handled gouge as he grinds the cutting edge on a 6-in. aluminum oxide wheel. Note how the tool's cutting edge rides directly on the blade and doesn't touch the grinder's tool rest, which is set for grinding skews and other straight tools.

Clamp an 8-in.-high rest to the arm and put your longest gouge in the support notch. Adjust the rest height and arm length so the wheel contacts the middle of the bevel. Now try your shortest gouge. You may have to compromise on the height to be able to work on both tools satisfactorily. When you have worked out the height, cut the rabbet and assemble the support. Note that when the handle is properly adjusted, the tool rides directly on the wheel and doesn't touch the grinder's tool rest at all. To grind square-edged gouges, first rotate the tool on its bevel. For a fingernail grind, you also have to push the tool up the wheel slightly as you roll the tool onto its side. After grinding, Zimmerman hones the tool with a medium India stone or soft Arkansas stone. *—D.B.*

backing paper and squeeze the blank to the faceplate for a few minutes with a handscrew.

Most of the hollowing was done with a deep-fluted gouge ground to a fingernail shape (bevel extends well back along the tool's sides). A ⅜-in. fingernail gouge was more maneuverable than a square-edge gouge for the deeply hollowed bowl shape I was working on. I could easily cut across the fibers of the wood on the bowl's outside and get a smooth surface by cutting from the bowl's small diameter to its large one. Hold the flute at about 10 o'clock, orienting the cutting edge at 45°. Remember to use your body, not your hands, to move the tool. Make a notch to begin the cut, then lower the handle and twist it around, adjusting the flute in the 9 to 12 o'clock range. The swinging motion of your body will arc the cutting edge to follow the curve of the bowl. For deep hollows, you can let the tool's shaft rub against the rim, using it as a fulcrum to cut deeper into the bowl. At first, I expected the edge to catch despite the rubbing bevel, but I kept adjusting the flute orientation and the shavings kept spewing out. I got so carried away that I cut through the bottom.

Zimmerman gets a remarkably clean finishing cut inside the bowl using one side of a ¼-in. deep-fluted, square-edged gouge with the edge held at about a 10° to 20° angle. During this operation, he held one hand lightly on the outside of the bowl to dampen vibration. You can also finish up with a scraper—scraping isn't a bad word in Zimmerman's shop. He feels it's important to enjoy and to feel comfortable with turning, and some people are just more comfortable with scraping. Zimmerman makes what he calls the slicing scrape with a roundnose scraper, which produces a fine shaving, not sawdust. To avoid the torn endgrain commonly associated with scraping, he cuts with the tool edge held at 45° to the wood's motion, so only one corner of the scraper contacts the tool rest.

Since I hadn't turned much in recent years, I think I went to the seminar with a fairly open mind and unpracticed hands, eager to develop new skills and perhaps to rekindle my interest in turning. Zimmerman's hands-on instruction helped me make sense of many things I had half-learned. Since the seminar I've been turning regularly again, and am finding that I'm cutting much faster and with greater accuracy than before, and producing crisper, more delicate pieces. And it's been fun. I couldn't ask for much more from any teacher. □

Dick Burrows is an associate editor of Fine Woodworking. *Russ Zimmerman's school is in Westminster, Vt., a small community in the southeastern part of the state. His address is RFD #3, Box 242, Putney, Vt. 05346. In addition to Myford lathes, he sells Permacel double-faced tape and Sorby tools, and publishes a chatty technical journal,* The Zimmerman Woodturning Letter.

Green Bowls

Turn unseasoned wood, dry it, then turn again

by Alan Stirt

A big problem in bowl turning is obtaining thick, wide, dry wood. You might be able to get 4-1/2 or 5-inch thick mahogany or 4-inch teak from an importer. In the Northeast you might find some 3 or 4-inch maple, birch or cherry at local mills. These planks usually contain numerous checks and splits. If they are sound, they will be more expensive than thinner material. If you want to turn a number of bowls, such sources will be quite frustrating in terms of cost and available species.

However, green (unseasoned) wood can readily be found and is often free. Even exotic woods are much cheaper when bought in the log. Working directly from the log gives you an opportunity to fit sizes and grain patterns to your own requirements, rather than accepting material that has been milled to a predetermined size. Green planks also offer advantages over dry wood. You can get larger sizes (the sawyer won't mind cutting extra-thick planks if he knows that *he* won't have to dry them), and the material will be in better condition.

In rural areas, logging waste — often containing the most figured wood — sawmill slabs and storm-damaged trees are usually free or sold cheaply. Firewood piles yield nice chunks of local hardwoods. Small local mills usually are glad to cut logs to whatever dimensions you want. Here in northern Vermont (in 1976), mills charge $40 to $50 per 1,000 board feet for milling logs you bring them. If you buy a log from the mill and have it cut, the cost is 20 to 30 cents per board foot. If the log is in good condition, such material is virtually check-free. Even in cities, green wood can be had from local tree-removal services and highway departments.

After you've found a supply of green wood, you have to dry it. One way is in planks or bowl-size blocks, but this is unlikely to produce perfect material. The easiest method is to turn the wood when it's green. Once the wood is in a bowl shape it dries much faster and with fewer defects than a solid chunk. You might start with a slab of lumber 4 or 6 inches thick, but if you turn the walls of the bowl down to an inch, it dries more like 4/4 stock. The analogy isn't exact because the grain orientation of the bowl isn't the same as that of milled lumber, but proper drying procedures minimize the differences. As the bowl dries it will warp and shrink, but once it is dry the walls are thick enough to be turned true again.

As an example of green turning, I'll show how to get a dry bowl from a green log of lignum vitae about 9 inches in diameter. It had been drying for about two years, but it was still quite wet. Similar procedures can be used for most hardwood species, both native and exotic.

First, cut about an inch off the end of the log to find check-free wood. If the log has been in the sun, it may be necessary to cut a series of thin slices to reach sound material. In some hardwoods small center checks run the whole length of the log, but these will be removed when trimming the block for the lathe. Next cut off a cross section as long as the diameter of the log, and rip this piece along the grain through the center of the log. If there are any center checks, make this second cut parallel to them and the saw kerf will often obliterate them. It is important to make sure the center of the tree —the pith—does not end up in your bowl as it will almost certainly split. Note any other checks and defects and plan your cuts to eliminate them from the final shape. Next, flatten the outside of each slab. This will be the bottom of the bowl. The flat surface will make the block safe to cut on the band saw. On the lignum vitae I roughly flattened the bottom with a 1-1/2-inch carving gouge, but these cuts can be made with a chain saw or a band saw. To cut down vibration and make turnings easier, I taper the sides of the block. I used the gouge but the easiest way is to saw a tapered circle. My band saw just doesn't have the capacity to make this cut.

The more you refine the shape with hand or power tools, the easier the initial turning will be. How far you go depends upon the size and species of your block of wood, the size and weight of your lathe, and your confidence and skill in using your tools. It's best to start with a balanced shape and discover how much unevenness you and your lathe can take. Even a small, out-of-balance piece can cause a lot of vibration.

First I turn the back of the bowl, with the face that was at the center of the log attached to the faceplate. Use long screws to grip the wet wood since the bowl will be absorbing a

Bowls turned from green wood by the author. Largest, 15 inches across, is of quilted, broad-leaf maple. Others (clockwise) are from zebrawood, white ash and cherry burl

From *Fine Woodworking* magazine (Summer 1976) 3:37-39

number of hard knocks in getting it true. Even if you don't usually wear a face shield when turning, it's important to wear one now. In the early stages chips will fly in all directions and some of them will be rather large.

Before turning on the lathe make sure the wood will not hit the ways or the tool rest. I start at a low speed and use a gouge, taking light cuts at first.

Don't try to decide the exact shape until all the rough spots are gone. Once the bowl is true, stop the lathe and carefully examine the wood. Note any defects which have to be removed, and interesting grain patterns to develop. The shape and the grain can be made to work together to create something more than just a bowl. On the lignum vitae bowl, I cut quite a bit off the bottom to ensure an interesting balance and pattern of heartwood and sapwood.

In shaping a bowl, I find the gouge to be the most efficient and enjoyable tool. The wood cuts cleanly and thick, curly shavings usually fly from it. Lignum vitae is an exception, preferring to come off as chips. Some woods, particularly butternut, are so soft and stringy when wet that they are hard to cut with anything but a gouge. A scraper just pushes the fibers around. To cut the straight foot, I use a 1/4-inch gouge with a slightly pointed nose.

When the contour of the bowl is done, flatten the bottom and make a pencil line to help reposition the faceplate.

Before remounting the bowl, I drill down to 1 inch from the bottom using a 1/2 or 1-inch bit. This gauges the depth and makes the gouge work easier. The faceplate can now be mounted on the bottom, using shorter screws because the wood will be running true. If you align two of the screws with the grain direction, the holes will probably remain in line during drying. Jot the screw size on the bowl for remounting later.

First I clean up the front, taking light cuts with the gouge. This can be a great help in reducing vibration, particularly if a chain saw was used to cut the log and the front is uneven. Now the bowl can be hollowed out. Because the wood is wet the tools stay cool and large amounts of wood can be removed before resharpening. I usually start at the center and work out toward the rim.

It's important to keep the thickness uniform throughout, so the bowl will dry evenly with less risk of checking. The thickness is very important in determining drying time, and a bowl turned down to 1/4 inch would dry very quickly with little chance of checking. However, it would distort more than a thicker bowl and when dry would be nearly impossible to turn truly round. For most native woods leave the walls and bottom about an inch thick. I gauge the thickness with calipers as the bowl nears completion, and examine it carefully for checks and knots. Checks present when the bowl is wet will get larger as it dries, and knots will often start checks that spread through the wood.

If you're satisfied with the condition of the wood, start the lathe and coat the bowl with a heavy layer of paste wax. I use Johnson's paste wax because it's cheap and I purchase it by the 12-pound case. Wax the bottom after removing the faceplate.

It's a good idea to rough-turn in an uninterrupted sequence. If you have to stop before the bowl is hollow, wax the wood to keep it from drying. I have had unwaxed pieces start checking in minutes in a heated shop.

Generally, the slower the drying the less risk of severe warping and checking; however, if the drying is too slow the wood may succumb to fungus and decay. And the slower the bowls dry, the more storage space the turner needs.

One controlling factor is the coating on the bowl. If left unsealed, the end grain will dry much faster than the rest. This can result in checking. Wax evens the drying rate and slows the whole process. So far I have used only paste wax. I'm sure any sealer that would adhere to wet wood would work to some extent. If I find that one layer of wax is not preventing checking I'll add more. The more layers of wax, the slower the drying and, up to a point, the less the chance of checking.

Each species of wood dries differently. In general, the higher the density of the wood, the longer it will take. But even within a single species the density can vary greatly. Sapwood will generally dry faster than heartwood and can cause extra distortion in bowls where both are present. Among domestic hardwoods, cherry and apple check easily while elm, walnut and butternut are excellent; in general, fruitwoods are more susceptible to checking than nutwoods. Ash may check within minutes.

This particular variety of lignum vitae proved to be very stable. Although I had to be very careful about checking, hardly any distortions occurred (by using many layers of wax and slow drying conditions I lost only one bowl out of 15 completed ones). These bowls I turned from 1/2-inch to 3/4-inch thick. I dry most native wood bowls, turned 1-inch thick, for about three months. I dried the lignum vitae bowls from six to twelve months, according to size and thickness.

You have great control over the drying environment, and the environment is crucial. Temperature, humidity and air circulation are the important factors. In the winter I never start bowls drying in a heated room. Usually I'll dry them in a spare room which stays around 45 or 50 degrees with moderate air circulation. After some weeks — the exact time depending upon the experience with wood of this species and grain formation — I move the bowls to a heated room.

A room which has good drying conditions during a period of high humidity can become an oven when the humidity drops sharply and stays down. Often the conditions can be changed just by opening a door, for increased circulation and faster drying, or closing it, to retard drying. If you want to be more scientific, you can outfit a room with temperature and humidity controls.

Once I found some 12 by 6-inch cherry bowls had checked during their first few days in my "normal" drying conditions. I dug out the checks with a gouge and rewaxed the bowls. Then I put them in my cellar which has high humidity. The bowls gradually dried without checking. However, they developed an unattractive blue-green stain from a fungus which thrives on high humidity. I completed the drying in a heated room and then finish-turned the bowls. The stain went deep into the end grain and was visible after finishing. I later dried the cherry in conditions that represented a compromise between my spare room and the cellar.

It pays to experiment with the facilities you have available; such experimentation should be a never-ending process. I have arrested checking by placing bowls in paper bags for a few weeks to choke off air circulation. Once you have an idea of the principles involved there are endless ways to deal with problems.

To determine when the bowls are at equilibrium with the

relative humidity and temperature of the surrounding air, weigh them periodically. When they stop losing weight they are dry. Under average conditions, most native woods rough-turned to a thickness of 1 inch will dry in about three months.

I should mention an alternative to the drying procedures I use. The green bowl can be soaked in a heated solution of PEG (polyethylene glycol 1000) before drying. The chemical replaces the water in the cells and prevents them from shrinking. I experimented with PEG a few years ago and was not satisfied with the results. The slight differences in appearance and finishing qualities mentioned by PEG's proponents were real differences to me. Also, I was having success with natural drying and saw no need to continue with PEG. It can be useful, however, because with it you can turn bowls that include the pith of the tree. I know one professional turner who's satisfied with the results and I'm sure there are more. For further information, contact the Forest Products Laboratory and Crane Creek Company, both in Madison, Wisconsin.

When your bowl is dry it can be finish-turned. First, plane the bottom flat. Before mounting, drill a hole to mark the finished depth. This will prevent turning through to the screws, which penetrate about 3/16 inch. I usually am able to use the same screw holes as in the rough turning, and I use the same length screw. Mount the bowl on the lathe and check to see that it clears the rest and the bed.

I true the outside first, with the lathe at low speed. I usually use a gouge but I found light cuts with a small round-nose scraper ideal for the lignum vitae, which is very hard when dry. A larger tool might have taken too big a bite and forced the bowl off the screws. I finished off the outside shape with a skew scraper.

At this point I usually sand the outside of the bowl. I turn most bowls relatively thin and when I am done hollowing, the walls vibrate. It's much easier to sand before hollowing, with little vibration. I start with 50 or 80 grit and work my way up to 220. I always wear a mask because the fine dust can be quite harmful.

Now I clean up the rim of the bowl with a gouge. Next I get the inside rim true and work my way down to the bottom, using a gouge and scraper. I advise against using the scraper on the sides of deep bowls because it can really make a mess of end grain. When I'm satisfied with the contours and thickness — measuring with calipers — I sand the inside of the bowl using the same grit sequence as on the outside. The bowl can now be hand-sanded, if desired, to remove circular scratches. Finish as you like.

The above procedures are only guidelines and can be adapted for almost any wood you'd care to turn. Exact methods of turning and drying should be worked out individually in one's own particular situation. I've had some failures and will have more in the future, but I've had a high rate of success. It is very satisfying to make a bowl when you control the whole process from log to finished form. □

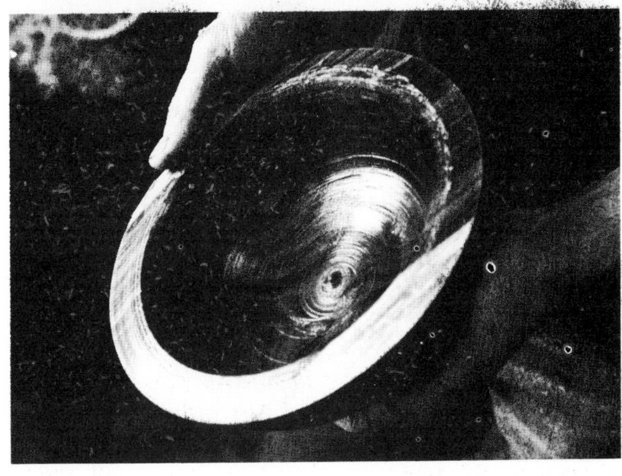

Before turning the back of a bowl cut from a green log (top photo), try to make it as round as possible. With the back turned, the faceplate is then attached to the foot, and the bowl is rough-turned. Then the whole bowl is liberally coated with paste wax (bottom photo) to control drying. When dry, the bowl is remounted and finish-turned.

Turnings Without Screw Holes

Make sectored-jaw faceplate chucks to hold the work

by E. Carroll Creitz

Traditional methods of mounting workpieces for faceplate turning leave something to be desired. Screw mounting leaves holes that must be plugged. Jam-fitting the workpiece to a recess in the faceplate requires critical fitting. Glue-and-cardboard mounting works well but the cleanup is time-consuming and irritating: The glue and paper fibers quickly ruin a sanding belt, which is costly these days. Washing off a water-soluble glue is a chore, and it raises the grain, which requires additional sanding before the finish can be applied. It seemed there had to be a better way.

The machinist's pot chuck (figure 2) appeared a likely candidate for adaptation to the wood lathe. The three jaws are sectors of a circle, and since pot chucks are usually furnished as blanks, the machinist can cut as many concentric gripping rings, of whatever diameter, as are required. Actuation is by a draw bolt in typical collet fashion: The chuck is pulled back into a tapered section, causing radial compression of the jaws. Pot chucks are not generally available for wood lathes, and they can't be made by most amateurs. But sectored jaws and the convenience of draw-bolt actuation are features worth having.

There are several ways to transmit longitudinal motion through 90°, the most practical of which is the bent lever. The evolution of a bent lever into a sectored faceplate is shown in figure 1. The chance of success in using a curved pivot point (figure 1c or 1d) seemed doubtful. So I temporarily shelved this design in favor of a bent lever whose movement would be provided by pressing a slotted disc into a dish-shaped cavity (figure 1e). The flexibility this would require could be provided by a thin, good-quality plywood, to which could be attached a jaw ring. The chuck I thus constructed is presented in figure 3. No dimensions are shown because they are a matter of convenience rather than necessity. Note, though, that the support plate must be thick enough that the screws attaching it to the faceplate will not interfere with recessing its face. Make the outside diameter of the jaw ring about ½ in. to ¾ in. larger than the diameter of the workpiece to be gripped. The thicker the stock from which the ring is cut, that is, the larger the distance between the base and top of the jaw, the larger the travel and the less the force on the workpiece.

I started with 8/4 stock to form jaws to hold a 4¾-in. diameter workpiece. This combination permitted a jaw travel of 3/16 in. I made the jaw ring about 1 in. wide to allow for a good solid glue joint (I used epoxy) between it and the ⅛-in. plywood, which forms the flexible member of the chuck. I centered this jaw assembly (ring and plywood disc) by first turning a flat-bottomed recess in the support plate, about ⅛ in. deep, and of a diameter slightly larger than the jaw assembly. Then, using a ¼-in. drill in the tailstock chuck, I drilled all the way through the support plate at its exact center. I unscrewed the faceplate from the headstock with the support plate still attached, inserted the jaw assembly into the recess, plywood side next to the support plate, and drilled a ¼-in. hole through the center of the plywood using the support plate as a guide. I next inserted a ¼-in. machine bolt, with washers, through both support plate and jaw assembly and tightened it to hold the jaw assembly in place.

The faceplate with its various attached parts was then returned to the lathe and the jaws turned. In the machinist's pot chuck, several concentric gripping surfaces can be used because they all move together and all exert the same force on the workpiece. Multiple gripping rings on a bent-lever chuck are not recommended because of the inverse relationship between motion and force. Accordingly, I turned only a single ring on this chuck. It is necesary to undercut the gripping ring a few degrees to form a circular dovetail of the required diameter. I then removed the jaw assembly from the support plate and remounted it in the recess with the plywood side out, so I could use the headstock index to mark for sawing the radial slots. I removed the jaw assembly and sawed the radial slots, leaving a 1⅜-in. dia. circle of solid plywood 1⅜ in. in diameter around the center hole. I deepened the recess in the support plate about 3/16 in. at the center hole, tapering to the outside edge of the recess.

A pressure plate is needed to press the plywood disc into this dished recess, so I cut a piece of ¼-in. Baltic birch plywood to fit inside the jaw ring and turned its inside face to about the same taper as the support-plate recess. I squared the hole in its center to accept the head of a ¼-in. x 20 carriage bolt, slipped the carriage bolt through and connected it to a draw bolt (¼ in. x 20 threaded rod) using a ½-in. piece of aluminum rod, drilled and tapped. A thick nut would do. The draw bolt passes through the hollow shaft of the headstock to the outboard side, where I attached a handwheel: a brass cone and a Lucite disc. The handwheel draws the pressure plate against the jaw assembly and tightens the jaws. Because the saw kerfs limit the radial motion of the jaws, I enlarged them with a handsaw.

I turned about 25 bowls and plates using this chuck, first screwing the blank to an ordinary faceplate and turning a foot to about the diameter of

Fig. 1: Evolution of chuck designs

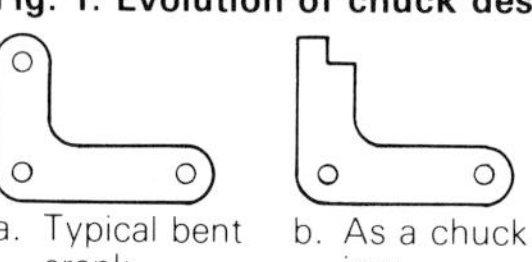

a. Typical bent crank b. As a chuck jaw

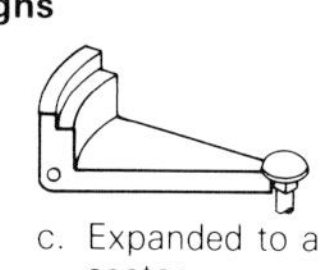

c. Expanded to a sector

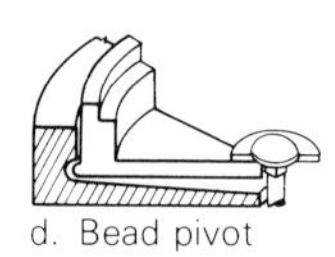

d. Bead pivot

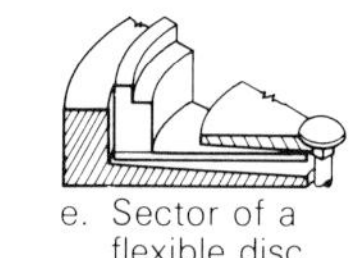

e. Sector of a flexible disc

Drawings: Robert Croston

Shop-built faceplate chucks hold the work without marring it and without large waste blocks attached. On the ways are dished-action chuck, left, and bead-pivot chuck, right. Expanding dished-action chuck, mounted on the lathe, is for bracelets.

Fig. 2: Machinist's pot chuck

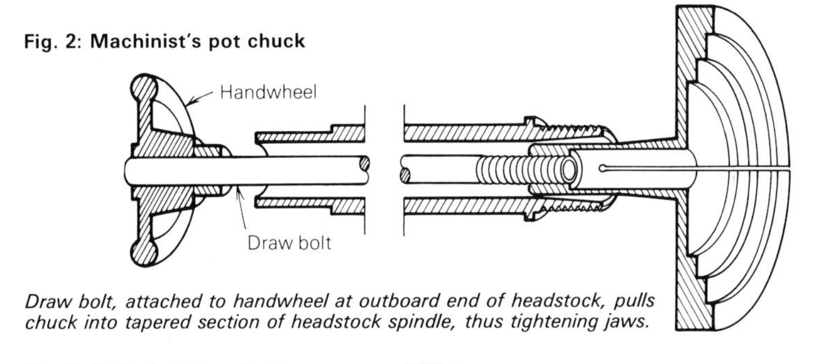

Draw bolt, attached to handwheel at outboard end of headstock, pulls chuck into tapered section of headstock spindle, thus tightening jaws.

Fig. 3: Dished-action chuck

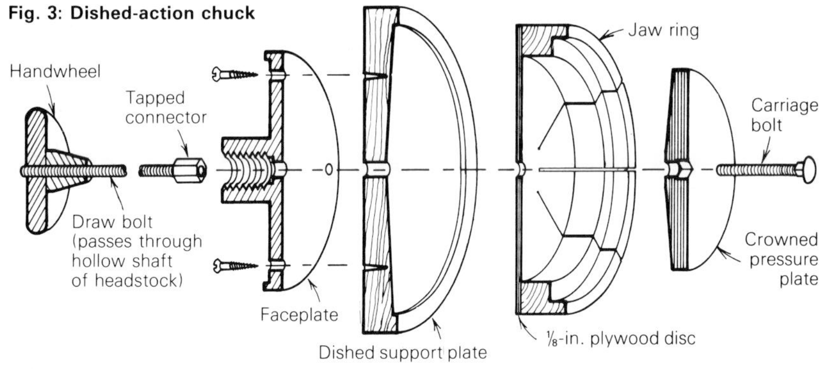

Pressure plate forces plywood disc to conform to dished surface of support plate, thus tightening jaws.

Fig. 4: Bead-pivot chuck

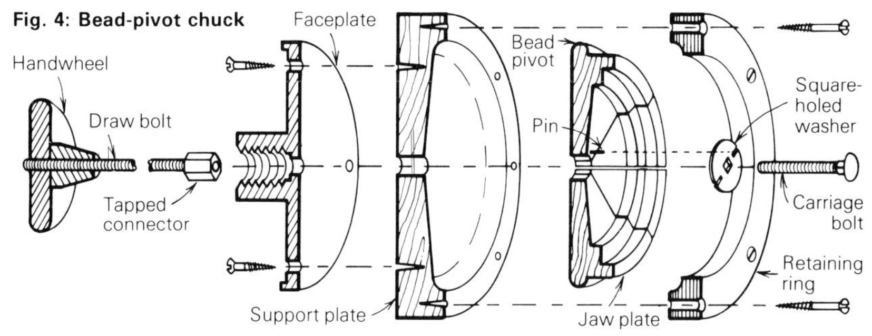

The eight separate sectors of the jaw plate pivot on a bead captured in the groove of the support-plate/retainer-ring assembly.

Fig. 5: Expanding, dished-action chuck

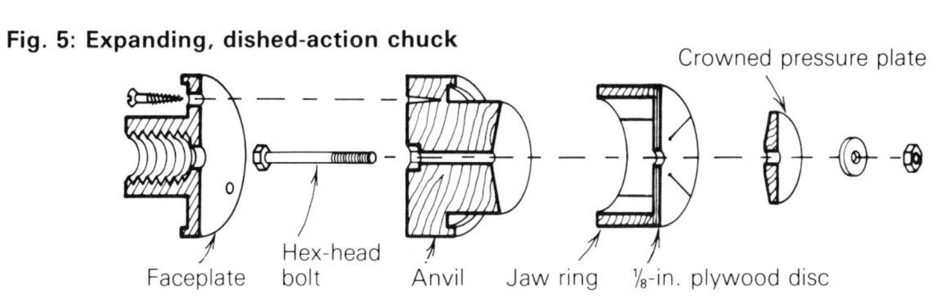

the jaws. The method was so successful that I made additional chucks for 3-in. and 6-in. workpieces. The holding power seems equal or superior to that of a screw center.

The success of the dished-action chucks prompted a return to the pivot-point type of bent-lever chuck, specifically the bead pivot (figure 1d). Thus I made the chuck shown in figure 4. Instead of the jaw ring being applied to a flexible disc and then segmented, I turned the jaw plate from one piece of hardwood, cutting a bead at the base that would pivot in a groove cut into the support plate and retainer ring. I left the bottom of the jaw-plate thick because sawing it into sectors would result in two sectors short-grained at their narrowest ends. I tried a one-piece support plate but found that a separate retainer ring alleviates tedious fitting of the jaw sectors. You can turn the jaw plate to an exact diameter, adding one saw-kerf width for the reduced diameter caused by sawing into sectors. Then the support plate can be turned, its recess dished, and the groove for the bead cut, all with the retainer ring attached. Unscrew the retainer ring and insert the jaw plate, sawn into sectors. The carriage bolt, with a square-holed washer, keeps the sectors from falling down. Pin the washer to two of the sectors to prevent it and the carriage bolt from turning.

Partly because of curiosity about how much punishment 1/8-in. ramin plywood would take, I made the expanding chuck in figure 5. I built it on a 3-in. faceplate, using a piece of 1 7/8-in. thick pine as the anvil against which the pressure plate deforms the plywood and expands the jaws. I left a shoulder on the anvil to help align the workpiece. Since the end of this chuck is left exposed, a nut and washer replace the handwheel and draw bolt of my other chucks. The machine bolt to which these attach is countersunk into a hexagonal hole in the anvil to prevent its turning when the nut is tightened. I made two of these chucks in diameters of 2 1/2 in. and 2 5/8 in. and with them have turned more than 70 bracelets. The surface of the jaws of the larger of them is getting slick and will soon need pieces of sandpaper glued onto its surfaces. □

Carroll Creitz, a retired research chemist, lives in Kensington, Md.

From *Fine Woodworking* magazine (November 1980) 25:82-83

If you plan ahead and have a thick enough blank, you can turn a bowl bottom of any shape without leaving a clue as to how it was attached to the lathe. The photo above shows the secret: grip the bowl via a small wooden plug, which you saw off after shaping.

The Bottom Line for Turned Bowls

Versatile chucking plug permits a variety of designs

by Wendell Smith

A woodturner can often improve a bowl just by realizing that its bottom is as important to its design as its rim or its overall shape. You may neglect the bottom, but if you submit a piece to a juried exhibition, you can be sure the judges won't. The best bowls don't reveal how they were attached to the lathe. But chucking procedures can interfere with good bowl design. The trick is to make the design you want, while leaving no trace of the method.

The simplest method of chucking is to screw the bowl bottom to a faceplate, but then the finished bottom must be at least as thick as the length of the screws penetrating into the wood. Unless you want a thick bottom and plugged screw holes, a more refined method is needed.

Other chucking methods can impose restrictions. Turning a recess into the bowl bottom for an expanding-collet chuck requires that you design a bowl with a rimmed bottom. If you turn a male plug on the bottom to fit some type of ring chuck, you'll have a footed bowl. On the other hand, facing off the bottom so that it is flat and can be attached to a scrap block with double-sided tape or with glue and paper does

From *Fine Woodworking* magazine (March 1984) 45:64-66

lead to a flat bottom.

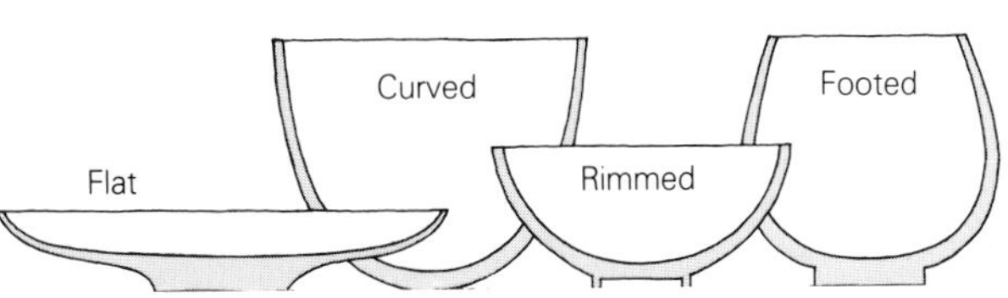

For green wood, I use a simple on-the-lathe method that lets me prepare and finish bottoms of any thickness, and have them flat or smoothly curved (with a slight flat spot to stand on), rimmed or footed, while maintaining complete freedom of design. This technique requires starting with wood thick enough to turn a ½-in. by 3-in. plug on the bowl bottom, using the plug to hold the wood on the lathe, and then removing the plug to finish up.

The plug is helpful when turning green bowls because the blank is solidly attached to a faceplate. You cannot use the glue-and-paper method because the glue will not stick to the wet wood. Balance is important with green stock because of its weight, and with the subsequently dried bowls because of their eccentricity.

For chucking dry wood, which is often not thick enough to allow for a chucking plug, I usually use glue and paper to fix the blank to a scrap block that is screwed to the faceplate. This requires a clean-up procedure based on a simple, off-the-lathe hand-scraping method (bottom right photo, p. 28).

The photographs illustrate my finishing method for removing the chucking plug from the base of an otherwise finished bowl about 5 in. high and 10 in. in diameter. This bowl was rough-turned green by first screwing a 3-in. faceplate to what would become the open (top) side of the blank, using ¾-in. #12 flat-head wood screws. With the faceplate mounted on the lathe, the bottom can be faced off with a deep-fluted gouge. Then, with the tool rest slightly below the center of the blank, use a parting tool to make a ½-in. deep shearing cut about 1½ in. from the blank's center. It is important to hold the parting-tool handle low when doing this, so the tool cuts, not scrapes. If your design calls for a footed bowl, cut deeper to leave a longer plug. With a gouge, shape the outside of the bowl to a rough form from the rim to the parting-tool groove. After the outside has been roughed out, remove the faceplate from the top of the blank and screw it to the plug. Although the bowl blank may not be perfectly recentered, it's unnecessary to true up the outside until the wood has dried. The inside of the bowl is turned using conventional methods. The photo sequence shows how to remove the plug, picking up after the bowl has been dried, re-turned and sanded.

1. The tailstock ring center holds the bowl, finish-sanded except for the chucking plug on its base, against a pressure plate made from a 1-in. by 13-in. hardwood disc. A groove in the plate holds the bowl's rim in place. Make a new groove for each size bowl, using a parting tool to size the groove until the rim of the bowl bears on either its inside edge or its outside edge. Now push the bowl onto the plate and bring the tailstock in. Before locking the tailstock, however, crank the ram far enough out to leave room for a small tool rest. At this point, any type of foot can be turned on the bowl. Here I chose a smooth, footless finish.

2. With a parting tool or a beading tool, make $\frac{1}{16}$-in. to ⅛-in. shearing cuts until the 3-in. plug is reduced to ¾ in. in diameter. Don't use too much force on the tool, or the bowl will slip. If the wood has a fancy figure, put the tool rest perpendicular to the lathe axis, then scrape away the plug with a small round-nose tool. A deep gouge could be used if the tool rest were lowered.

3. After reducing the plug to a ¾-in. diameter, smooth and flatten the bowl bottom using a straight-across or right-skew scraper, followed by sanding. Then, with the tool rest parallel to the lathe axis, use a thin parting tool to make a shearing cut about $\frac{1}{16}$ in. from the bowl's base. Before cutting deeply, widen the cut slightly on the right. Keep the cut wider than the tool as you cut into the plug, to reduce resistance to cutting and to keep from breaking the bridge. I find it best to rotate the parting tool slightly clockwise and counterclockwise while cutting, as though cutting a small bead. The small $\frac{1}{16}$-in. platform of waste wood left between the base of the bowl and the bridge prevents the parting tool from tearing wood fibers on the bowl bottom.

4. With the lathe off, I use a Japanese *dozuki* saw to cut the small bridge between the bowl and the plug. Before sawing, pull the tailstock ram back slightly to take pressure off the bridge. The masking tape protects the bowl bottom from the saw. For a full view of the work at this stage, see p. 26.

5. Place the bowl rim-down in a right-angle stop-block jig as shown at right and remove whatever waste remains by slicing cross-grain with a bench chisel, held bevel-up. The tape prevents the chisel from damaging the finished base. Finally, scrape the center of the base with the grain, then hand-sand.

Glue-and-paper chucks—To remove a glue-and-paper chucking block, place the completed bowl face-down on a towel and tap in an old plane blade, bevel-up. Insert the blade between the two bottom plies of the plywood, rather than between the block and the bowl, to prevent damage. Lift the blade end to lever off most of the block. The remaining waste can be pared off with a bench chisel, used bevel-up.

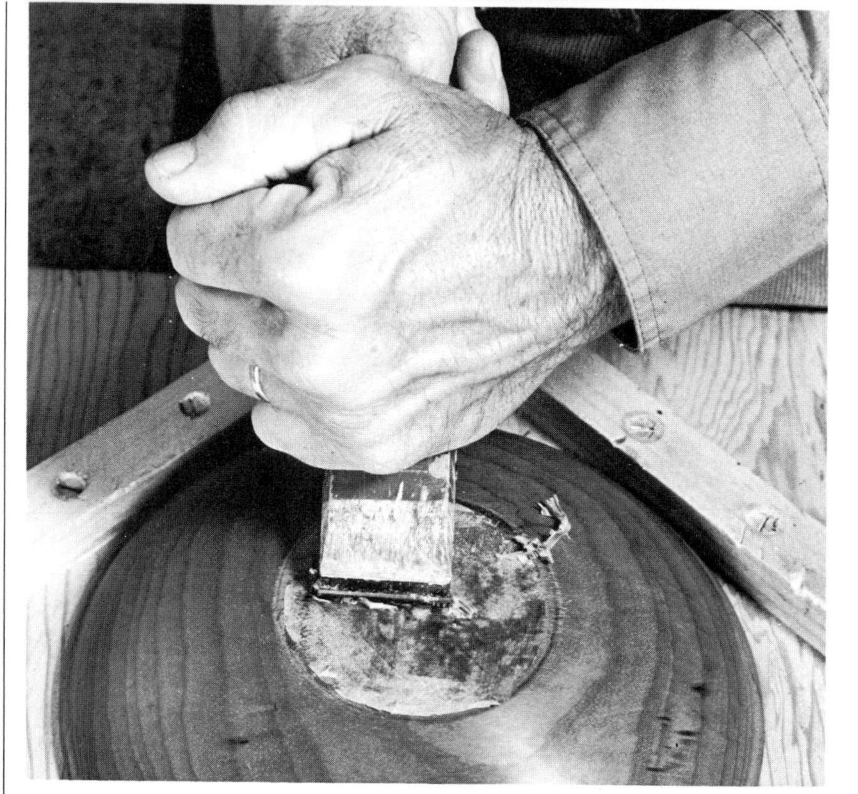

Remove the final traces of the glue-and-paper joint with a $1\frac{1}{2}$-in. paint scraper, then follow up with sandpaper. The secret to using these scrapers is to leave a burr on the cutting edge when sharpening it on a grinder. □

Wendell Smith, who lives in Fairport, N.Y., is a chemist in the Kodak Research Labs. Photos by the author.

Turning a Matched Set of Bowls

Patternmakers' tricks for consistent shapes

by Arthur F. Sherry

Getting the most from an outstanding piece of wood by making a one-of-a-kind bowl is part of the woodturner's art. But turning a good matched set of bowls can be an equal challenge, calling for careful planning and execution. A matched set, to me, means consistent shape more than anything else. Bowls can be made of different woods, or be inlaid with elaborate designs. Yet if their shapes are the same, we instinctively know they belong together. Here are some patternmakers' tricks and templates that will help you turn a series of bowls, or almost anything else, exactly alike.

Wood never stops moving as its moisture content changes, of course. Plan to use dry wood, or your bowls will become oval after they have been turned. I frequently rough-turn bowls, then let them dry for a few days to stabilize before I finish turning. I've found species such as mahogany and walnut to be particularly stable, but you can apply these techniques to more highly figured species, too.

Start by designing the shape on paper. Then transfer the layout to a squared piece of ⅛-in. plywood (figure 1). Lay out the centerline of the bowl, marked C/L. Then, with a knife, scribe lines for the top and for the bottom of the bowl, perpendicular to the centerline. Draw the cross section of one half of the bowl on the template, and scribe rim lines (parallel to the centerline) to mark the outside diameter of the bowl. Notice that the side of the rim should be left at least ⅛ in. thick, so that after the inside has been turned you can mount the bowl as shown in figure 4, for turning the outside.

Cut out, file and sand the template to shape. If I am making more than a few bowls, I copy this template onto another piece of plywood and use the master only for the final fit. I never touch the master to the spinning bowl.

To turn the inside of the bowl, screw the blank (bandsawn round) to a faceplate and mount it on the lathe. Turn the block to the final height of the bowl, plus $\frac{1}{64}$ in. for final sanding. Next turn the diameter, and stop to check it with both a square (so that the side is perpendicular to the face) and a ruler. I measure with a ruler, as shown in figure 2, instead of using calipers, because calipers have a tendency to give a little—a ruler is more accurate. First, mark the center of the blank while the bowl is turning, then stop the lathe and hold the ruler so it crosses the center point. If you stop the end of the ruler against a small wooden block held against the side of the bowl, the ruler will line up exactly with the edge of the rim.

You can hollow the inside of the bowl quickly at first, checking your progress with a template copy held against the spinning work. But stop the work often to check the fit as you approach the final form, as shown in figure 3. Keep in mind that the centerline of the template must end up at the center of the bowl, and that both rim lines on the template must line up with the rim of the bowl.

Stop turning when the inside of the bowl is about $\frac{1}{16}$ in.

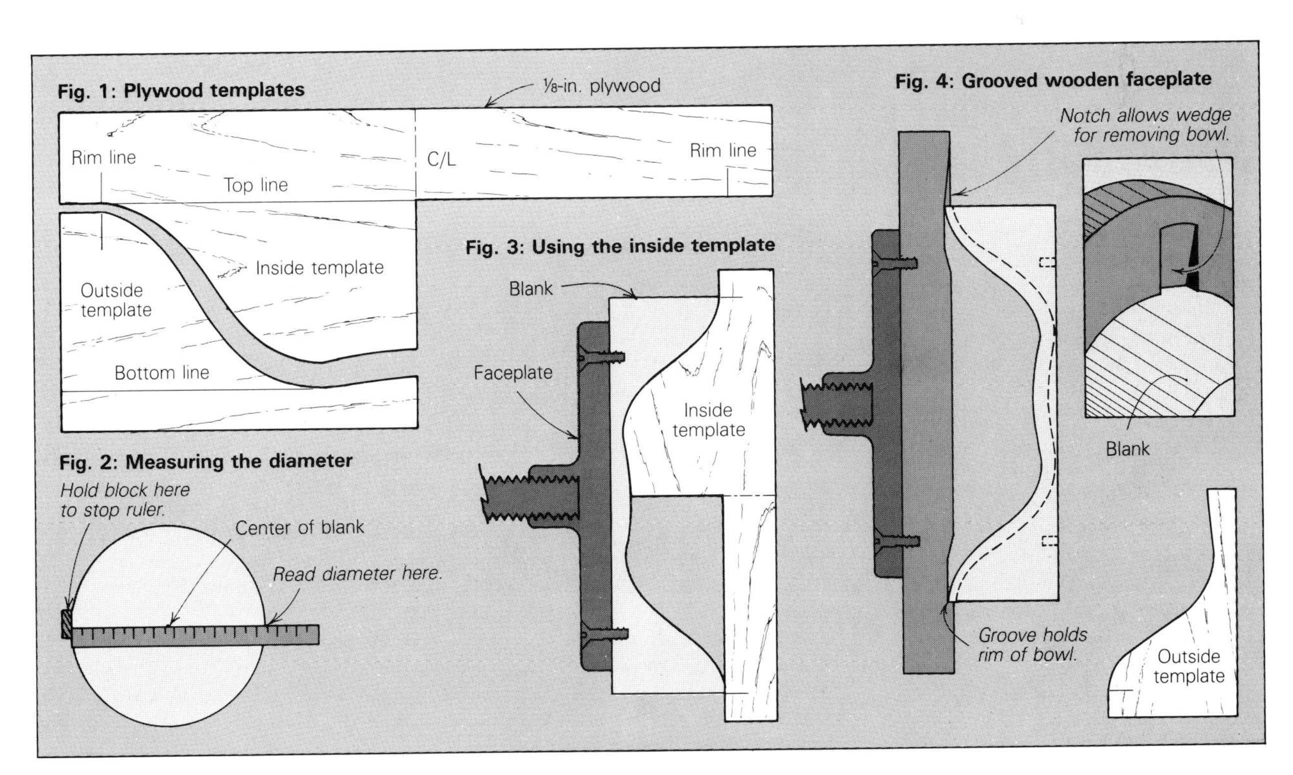

From *Fine Woodworking* magazine (January 1983) 38:70-72

full of these final marks, then switch to the master template. Rub the edge of the template with a little chalk or a crayon. Stop the lathe and rock the master back and forth in the bowl, gently transferring chalk to the high spots. Carefully turn away the marks, stopping and checking after every cut, until the master deposits an even spread of chalk along the profile of the bowl, but still about $\frac{1}{64}$ in. full of the reference points. Sand down to the line, using from 180-grit to 360-grit sandpaper, but leave the outside rim square so it can be mounted in the next step. Take every bowl in the set to this stage before proceeding.

To turn the outsides, begin by scribing a line that shows the location of the bottom of the rim. This will be the reference line for the outside template. Then check the diameters of all the bowls. There's always some slight difference, sometimes due to wood movement, sometimes to that last pass with the sandpaper. Select the smallest and turn a shallow groove in a wooden faceplate so that the rim of this bowl fits tightly (figure 4). There is no room for error here. Cut the opening with a skew chisel until its outside is slightly smaller than the rim of the bowl. Then turn the chisel over and rub, rather than cut, the last few thousandths away, until the bowl fits tightly and is difficult to remove.

We will hold the bowl in with a few tiny spots of glue, then use little softwood wedges or give it a light rap with a hammer to pop it out of the groove after the turning is done. Make some shallow notches in the faceplate before you glue the bowl in, so you will be able to get the wedges beneath the bowl's rim.

Mark a circle on the blank, approximately the size of the bottom of the bowl. Then turn the underside of the bowl using the center and rim line as guides, testing as before, until the chalk shows no more high spots. Switch to the master template and finish the bowl. Remove it with the wedges.

Enlarge the groove in the faceplate if necessary, to fit the next larger bowl, then repeat the process. When all the bowls have been turned, I use files and a piece of sandpaper glued firmly to a block to shape the rims, and I check the curve of the outside edge with radius gauges (standard sheet-metal templates). Because matched bowls are usually used for food, I finish them with a non-toxic finish such as Constantine's Wood Bowl Seal, or a vegetable oil.

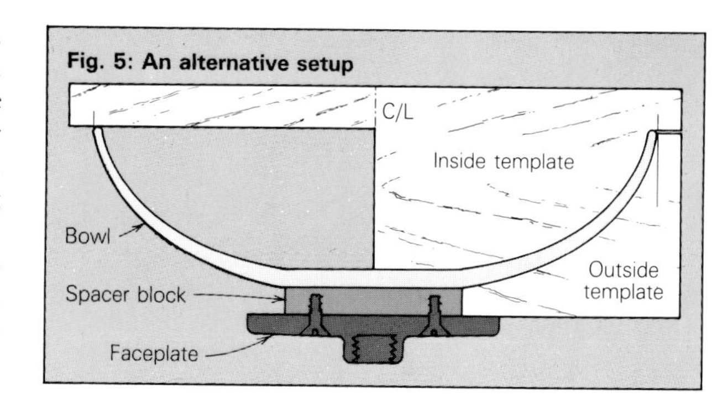

Once you understand how templates work, you can vary their use. A single-mounting setup that lets you work on the rim more easily is shown in figure 5. Its drawback is that the bottom cannot be as easily shaped. You have to glue uniform spacer blocks to the blanks and allow for their thickness in laying out the design on the template.

You can also take the guesswork out of long turnings. Just use several smaller templates along the length of the turning, each with its own set of reference points on the straight sections. If your lathe allows you to remove and replace your turnings accurately, do each step on all the turnings before proceeding to the next; if it doesn't, make a list of the steps so you can repeat them exactly in order.

As in all woodworking, accuracy on the lathe is as much a state of mind as it is a procedure. Templates will show you when you have gone far enough, but only your skill as a turner will prevent you from going too far. □

Arthur F. Sherry, of New York City, has completed four years as an apprentice patternmaker, and is now a partner in South Family Furniture, making custom furnishings.

Walnut-oil finish is safe for food

by Antoine Capet

A few hundred years ago, rubbed oil made do as a finish for everything from the cogs in wooden clocks to the gear on old sailing ships. When we think of rubbed oil, most of us probably think first of linseed oil, which is the most prevalent of the traditional oils, at least for outdoor items such as gateways and for seafaring. Yet many of us shy away from using it on bowls or other receptacles for holding food because modern, fast-drying linseed oils usually have poisonous chemical additives. The odor of linseed oil, also, while pleasant on a tool handle or in an artist's studio, quickly takes away one's appetite.

There are several other oils that can be used instead. A classic book on finishing, *Lexique du Peintre en Batiment,* by Le Moniteur de la Peinture (Paris-Liège, 1935-36), lists other natural oils and their drying capabilities:

Oils, high drying capability—linseed, poppy, tung, walnut, hemp, sunflower.

Oils, moderate drying capability—colza, soya.

Oils, no drying capability—olive, peanut, almond, castor, grape pips.

Some of these rule themselves out. Tung oil is not edible, poppy oil (from artists' supply stores) is exorbitantly priced, and hemp oil is unobtainable these days. I've left the moderate-drying oils alone, because they seem to have no advantages. Olive oil, often mentioned as a salad bowl finish, has the drawback of never drying.

Walnut oil, though, the traditional French furniture polishing oil, deserves a closer look. It is not only edible, it is delicious. And it can be bought in health food stores and specialty food shops at a price that compares favorably with modern finishing oils.

Walnut oil's pleasant odor and non-toxic qualities are in sharp contrast to some other finishes I've tried. One commercial salad-bowl finish, though certified safe for bowls, has a strong smell of petroleum distillate that persists for a long time. Another "certified safe" finish requires that you wait 30 days before actually using the object.

There are additional advantages to using walnut oil on functional objects. Quick and soft finishes, waxes for instance, poorly resist spilled coffee, wet hands and damp fruit. Walnut oil takes these things in stride. Many hard-film finishes can chip, crack or peel away, but walnut oil penetrates deeply, and will conform to a dent without losing its ability to protect against moisture.

What then is wrong with it? Walnut oil requires time to build into a decent finish, ruling it out in a cabinetmaking shop that seeks a quick, high gloss. But for many of us, making things for our own pleasure, this is not so important.

I use walnut oil without a sealer, because it accentuates the figure best when allowed to penetrate deeply into the wood. I made some tests on fenceposts and found little difference in its drying time (about the same as raw linseed) whether I added small quantities of drier or not. But I found a pronounced acceleration when the oil was applied hot. It can be heated for a few minutes in a saucepan, about one-quarter full, until fumes begin to thicken. There's always a danger of fire, of course, but people safely fry with hot oil every day.

The smoother the surface texture, the less oil the object will absorb, and the faster it will shine. You can use a paintbrush to apply the oil, but it's better if you dip the wood, because the end grain will gulp up vast quantities, ensuring protection against future checking. Work the oil into the wood, rubbing surplus from the sides around into the end grain. After a day or so, polish it at high speed on the lathe.

I usually wait two weeks before giving the wood a second application, then two or three months, then a year or more between further treatments. □

Antoine Capet teaches contemporary British social history at the University of Rouen, in France.

Turning goblets

by J.H. Habermann

Turning a goblet presents a few problems, but once you see your way around them, the job becomes easy.

Design: Some turners have enough confidence to let the shape evolve as they work, but I generally pick a shape I like—a favorite wineglass, for instance—and trace its outlines with a contour tracer, a pencil mounted on a base that follows the profile of the glass to make an outside pattern.

Once you have traced the outline and allowed ¼ in. for the walls, you can plan to drill out most of the inside with Forstner or multi-spur bits chucked in the tailstock. This will save wear and tear on your turning tools as you remove the difficult end grain. Determine from your own pattern which size bits to use and how deep to drill, as shown in figure 1. Use calipers to measure the diameter of the glass at several points, then mark these dimensions on a template made from hardboard.

Fig. 1: Hardboard template, drilling guide

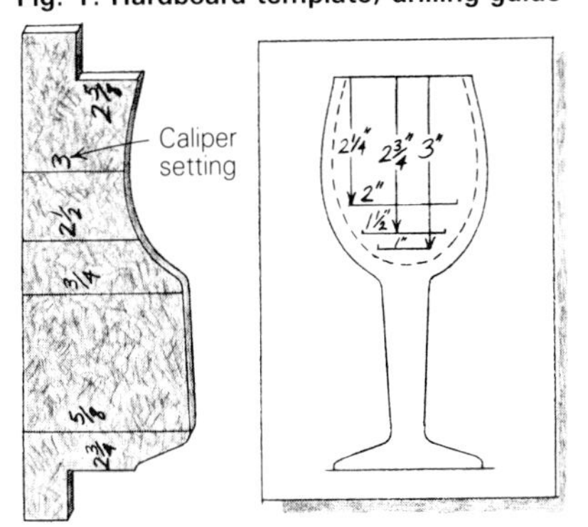

Fig. 2: Stub-tenon mounting

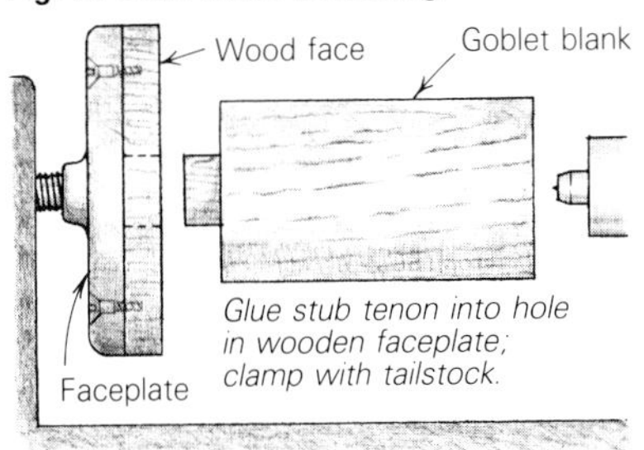

Fig. 3: Shopmade scraping tool

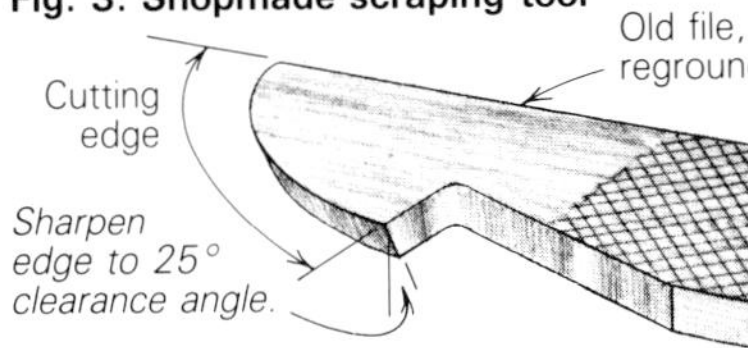

Fig. 4: Steadying the blank

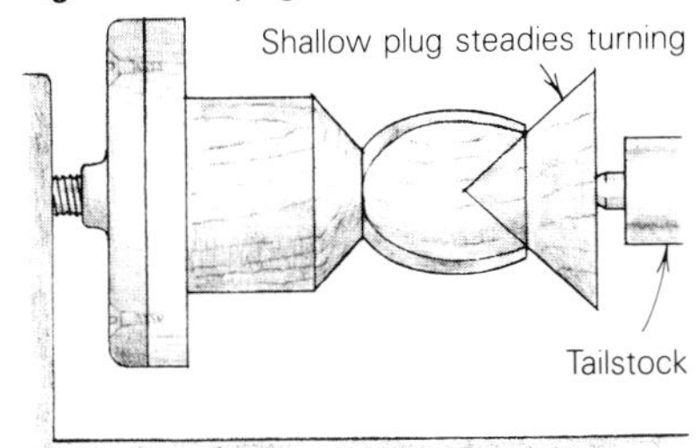

Wood: You can make almost anything into a goblet. Even native "weed trees" such as sumac work well. Goblets don't require scarce, chunky turning blocks; offcuts from furniture wood can be used. You can laminate thin stock either vertically or horizontally to get enough mass—just don't try to glue end grain. And leave some extra length. This allows you to turn a stub tenon for mounting the work in a wooden faceplate. You wouldn't want to waste this precious depth on a bowlturning blank, but here it doesn't matter.

Turning: Glue the stub tenon into the faceplate, aligning and clamping it with your tailstock, as in figure 2. When it is dry, rough-turn the outside of the top of the goblet, referring to the caliper sizes on the template, but do not turn the narrow stem yet. Chuck the appropriate drill bits in the tailstock, and drill out as much of the inside as you can.

True the inside by hand. Turners generally look down on scraping tools, but there's a lot of end grain in a goblet, and this is where scrapers excel. Commercial side-cutting scrapers such as Sorby's work well, but you can make your own by grinding old files, as shown in figure 3. I use my thumb and fingers as a thickness gauge.

When the inside of the goblet is true, insert a shallow plug (figure 4) and draw up the tailstock for stability. This will save you a lot of blown-up goblets as you turn the stem.

Using a combination of calipers and the template, turn the stem and clean up the shape. When I am duplicating a series, I make a full template that slips over the entire goblet while the lathe is stopped. This solves the problem of registering a half-template.

Before final-sanding, partially part off the base. Point the parting tool slightly toward the tailstock. This will give you a concave bottom that will be more stable.

Sanding and finishing: You will get far fewer circular scratches if you use an orbital sander to sand the work while it is turning. Work down to 280-grit, reversing the lathe once in a while if you can, and finish up with steel wool and a final polish with a handful of chips.

You will need to seal the goblet if you plan to use it. I have had great success with John Harra's DPS (deep penetrating sealer), which plasticizes in about a week. If you want a higher gloss, you can finish over this with a natural drying oil, or you can use commercial salad-bowl finishes. □

J.H. Habermann lives in Joplin, Mo.

Turning Tools That Cut

A book from Sweden favors some old tools

by James Rudstrom

My father's shop was equipped with all sorts of woodworking tools, and I was gradually introduced to their secrets. We had quite a number of machines, but no lathe. I guess one just didn't seem necessary for building houses and cabinets, and our economy permitted only necessities.

I was in ninth grade when I first tried turning. I don't recall being taught anything other than scraping, with the exception of attempting to make V-cuts with a rather dull skew. Several of my would-be masterpieces were ruined when the skew, instead of making its path toward the bottom of the V, skittered in the opposite direction, scarring everything in its path. That was my first experience with the cutting method of turning, although it took 20 years or so before I understood what I was up to.

I don't remember if it was because of my shop, his kids or his interest in English that I got to know my neighbor, woodworker and crafts teacher Wille Sundqvist. In any case, we hit it off well from the very beginning. In 1978, while he was teaching an evening course in my shop, Sundqvist mentioned a book that he was writing on turning. We had touched on the subject before, but not at length or in depth, probably for the same reason that I hadn't gotten to turning before—I still didn't have a lathe in my shop.

Sundqvist and another turner, Bengt Gustafsson, have published their book, *Träsvarvning enligt skärmetoden* (LTs förlag, Stockholm, 1981). *Turning, The Cutting Method,* as its title translates, is not a large volume, but it covers many aspects of lathe work that have been more or less forgotten. The book touches on lathe construction, tool design, tool care, safety, the turning of cylinders and profiles, bowlturning, green-wood turning and more. It is well illustrated with photos and drawings. Unfortunately, it is not available in English yet; the authors are exploring the possibility of U.S. publication. Until then, here are some of the book's most interesting points, for English-speaking woodworkers.

The cutting method presented by Sundqvist and Gustafsson would definitely be suited for efficient use on treadle lathes and other low-tech machines. High RPM are unnecessary; in fact, undesirable. In most cases, the tools advocated are almost opposites to those in the "long-and-strong" school. "Short, sharp and sensitive" is the name of this game. If you're interested in trying the approach, you'll find that some of these traditional tools are hard to come by. You may find some treasures at auctions and flea markets. But most of us will have to modify standard tools, make our own or find someone who can. Even with the right tools, this kind of turning requires industrious practice, but in return there's little dust, surfaces that require little or no sanding, and the good feeling when fine shavings reel off the workpiece.

Tool rests—Many of the standard and auxiliary tool rests on the market maintain too much distance between their point of contact and the tool's cutting edge. When you use a short, light tool, it is essential that the tool be supported as close to the cutting edge as possible. The usual type of rest, which has a flat for the bearing surface (figure 1), will shift this contact point away from the stock when the tool is positioned at the steep angles often used when cutting, especially in faceplate turning. The cleanest cut is usually obtained at a steep angle, very close to the threshold of kickback. Resting short finishing tools at point B in figure 1 will obviously increase the risks of chatter and kickback. Sundqvist and Gustafsson suggest the solution in figure 2. Notice the removable pins that prevent tools from slipping into the chuck.

Fig. 1: Standard tool rest

Tool rests topped with an inclined ledge are unsuitable for the cutting method. Supporting a tool at point B puts too much distance between the fulcrum and the cutting edge.

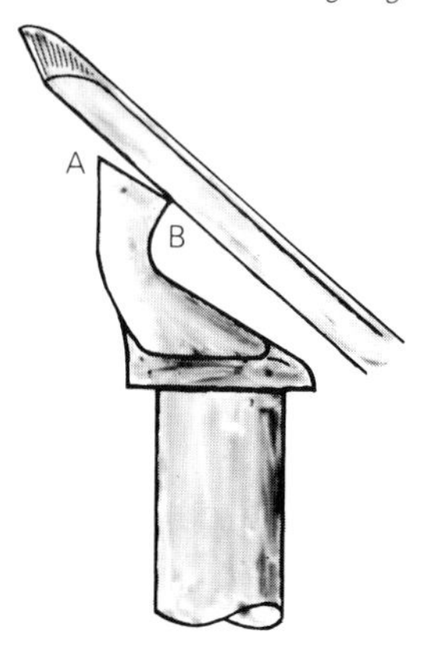

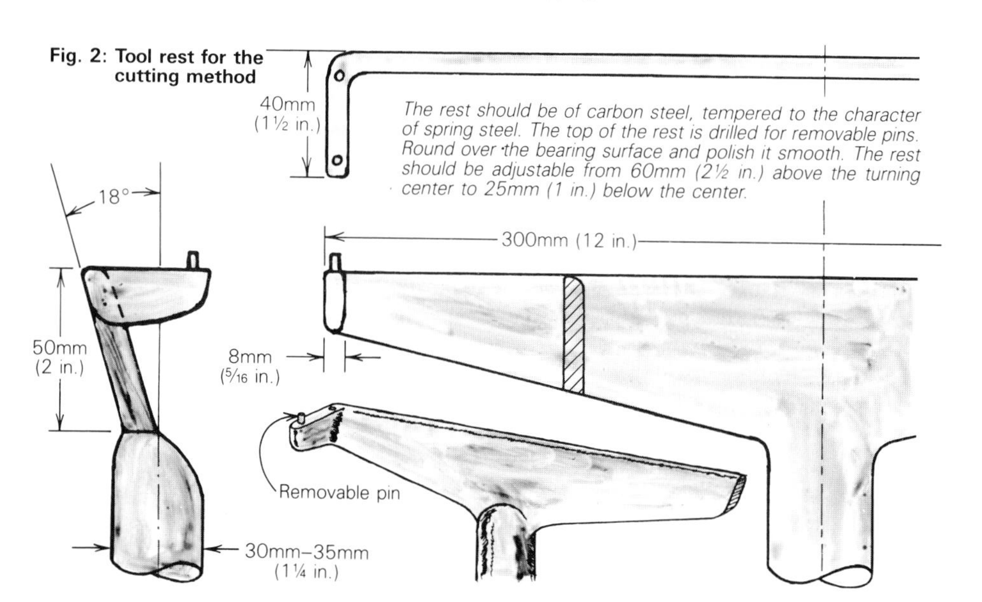

Fig. 2: Tool rest for the cutting method

The rest should be of carbon steel, tempered to the character of spring steel. The top of the rest is drilled for removable pins. Round over the bearing surface and polish it smooth. The rest should be adjustable from 60mm (2½ in.) above the turning center to 25mm (1 in.) below the center.

Photos and drawings adapted from *Trasvarvning enligt skärmetoden*

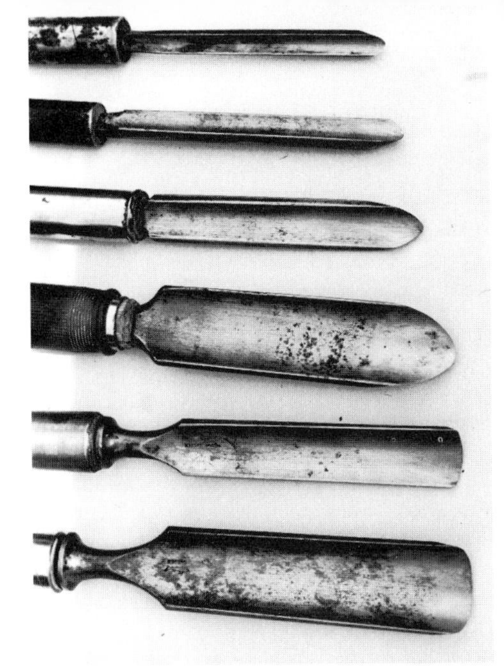

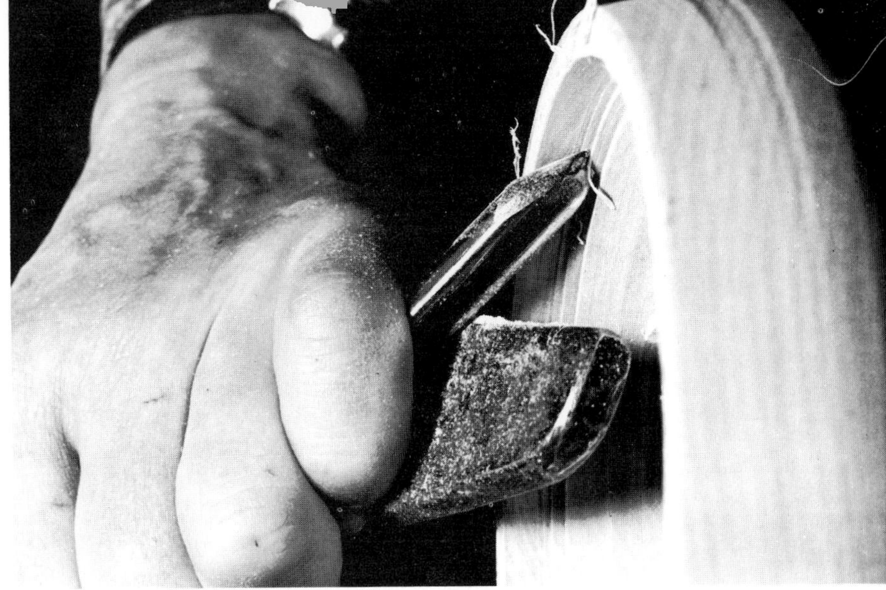

The tools at left were made by regrinding standard carvers' and turning gouges. Smoothing gouges (upper four) have an edge more curved than the edge of roughing gouges (lower two). Above, a gouge smooths the inside of a bowl.

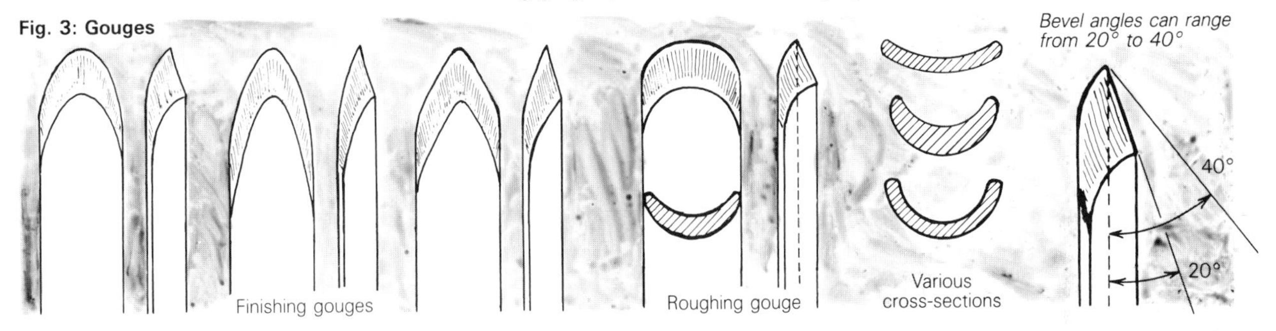

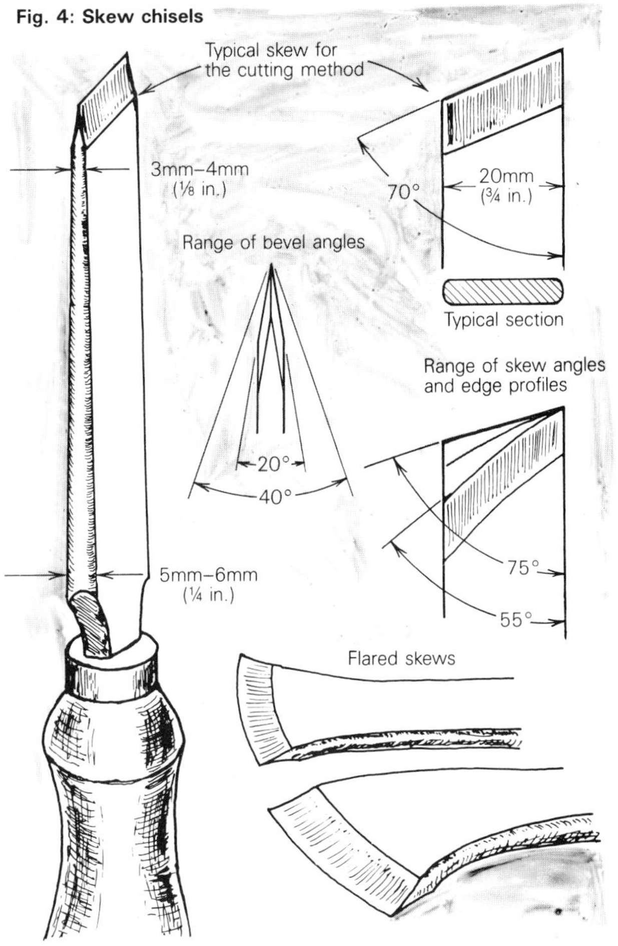

Gouges—Two main types of gouges are mentioned: those used for roughing out and those used for forming and finishing. Examples of different edge shapes are shown in figure 3. Note that the angle of the bevel can vary between 20° and 40°, following the general rule of a greater angle for harder or more figured wood. Many of the gouges on the market are too long for finishing work. The authors prescribe using imagination in regrinding standard turning gouges and in experimenting with the great variety of carvers' gouges available. In some cases, you might want to forge your own.

Skew chisels—The skew is perhaps the most versatile of all the turning tools used to cut. In the hands of a skilled turner, the skew can make profiles appear as if by magic. The short and rather thin skews recommended here should have a taper, being thicker toward the handle, as shown in figure 4, to lessen vibration. The corners along the length of the tool are rounded and polished for a better feel in the hand and to facilitate a light, gliding movement over the tool rest. The normal angle of the skew should be 70°, although different angles and variations from straight edges are common. Hand-forged skews were often flared at the edge and had long bevels, making them thin and liable to overheat on modern lathes with high RPM, especially when turning harder wood. Bevels otherwise are between 20° and 40°, as with gouges.

Turning hooks—Among the more valuable subjects in the book is that of turning hooks or hollowing hooks. These low-speed tools are near-forgotten things of the past. In essence, the turning hook is a gouge with its cutting edge oriented 90° to the length of the tool. It can have a double edge or a

From *Fine Woodworking* magazine (May 1983) 40:92-94

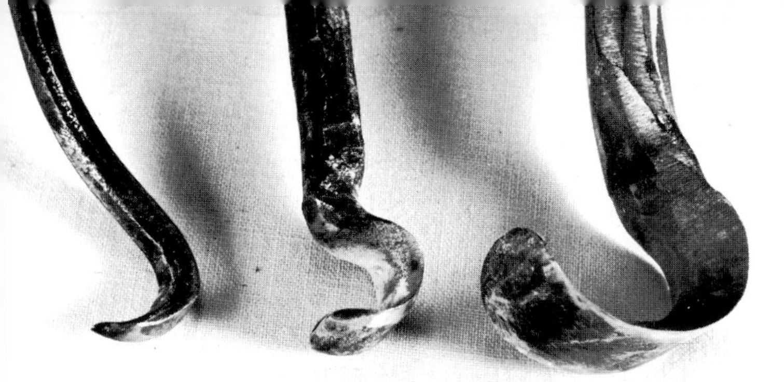

Turning hooks are actually gouges with the cutting edge turned 90°. The tool at left in the photo above was forged from a pitchfork tine. In the photo at right, a hook hollows a bowl.

Fig. 5: Turning hooks

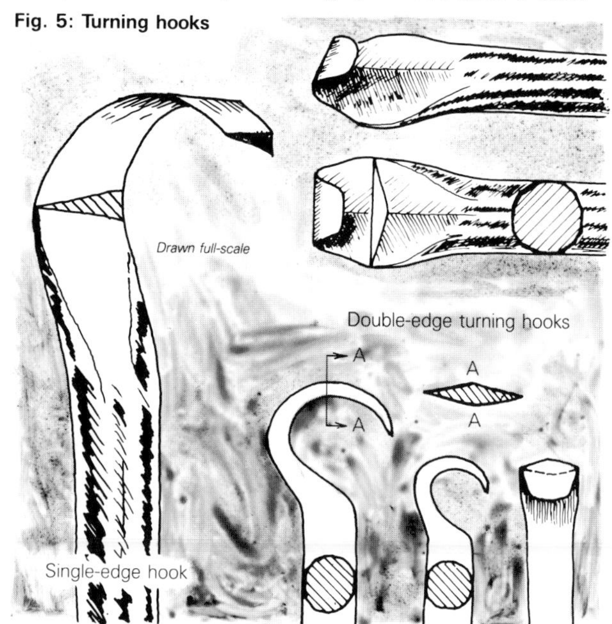

Fig. 6: Parting tools

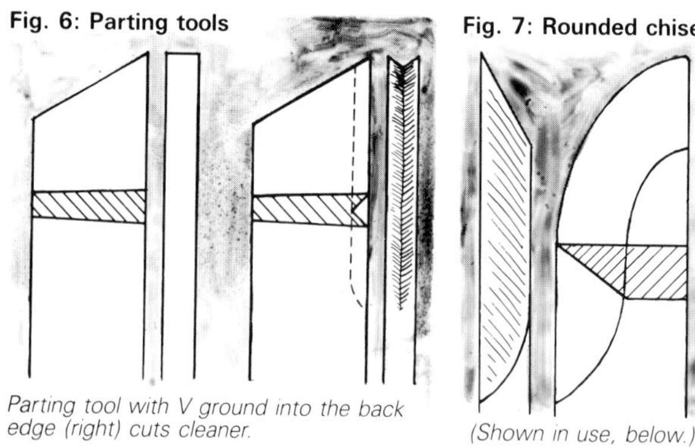

Parting tool with V ground into the back edge (right) cuts cleaner.

Fig. 7: Rounded chisel

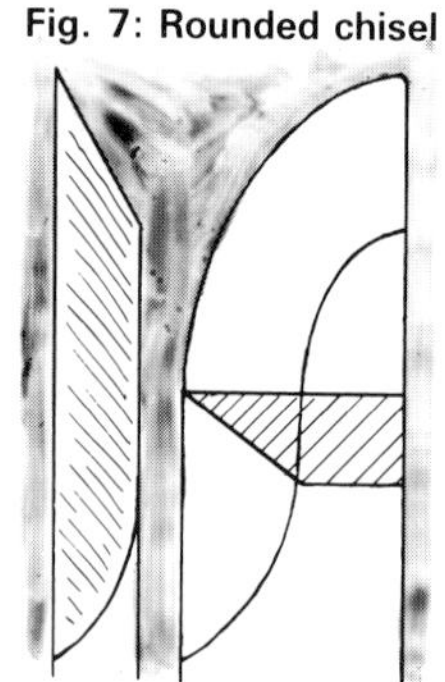
(Shown in use, below.)

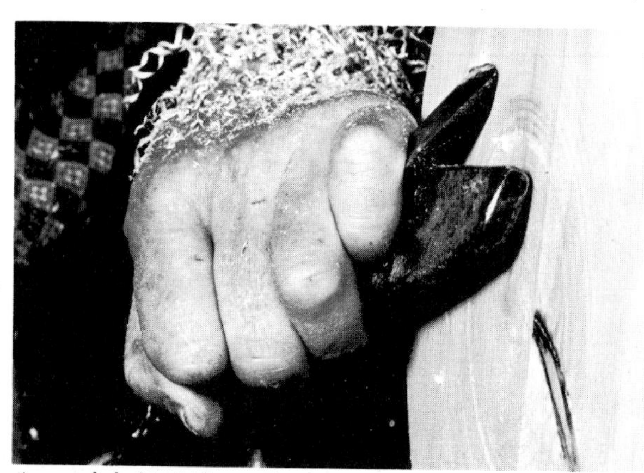

A rounded chisel (figure 7) planes surfaces in faceplate turning. Its bevel rides the work.

single edge, as shown in figure 5 and the photos above.

The turning hook cuts with the grain when working cross-grain on a faceplate, starting from the outside and cutting toward the middle. Cuts in end grain are the opposite, starting from a bored hole in the middle and working outward. To work in tight places, some turning hooks have the bevel ground on the inside rather than the outside. Turning hooks more than ⅝ in. across are best for softer woods. Long handles are also recommended for larger hooks, to help prevent chatter and kickback. Sometimes a bent or an angled handle facilitates holding the tool at an efficient, safe edge angle.

Other tools—A parting tool with a flat back will do; one with a V ground into the back (figure 6), however, will produce a cleaner cut. The blade is thinner toward the bottom edge to provide clearance. The tool needs plenty of metal beneath the cutting edge, and a long, sturdy handle to help avoid chatter and kickback.

For faceplate turning, the book recommends a rounded chisel (figure 7), which can be fashioned from a scraping tool.

Tool length—It is difficult to give definite rules for turning-tool length. Generally, large skews, roughing gouges and parting tools are about 450mm (18 in.) in overall length. Tools used for making profiles and doing other finishing work should be very short, 300mm (12 in.) or less. It should be possible to move these tools in front of you without having to hold your breath and pull in your stomach. Experienced turners work with smooth movements, switching hands when necessary. □

James Rudstrom is a school psychologist in Vilhelmina, Sweden.

Bent Bowl Gouges

Reforge your tools for finish-turning

by Douglas Owen

Owen turned this elm-burl bowl green, finishing it with one of his bent gouges. He says, 'It is difficult to season burr woods without all the little knots cracking badly. I couldn't sand the wet wood on the lathe, and I didn't want to sand after the wood had dried. I wanted to keep the wrinkled surface that comes from the wood's drying and warping, the more the better in this case. So I had to get a fine finish directly from the tool. I got what I wanted from one of the first bent gouges I made.'

The idea of tools bent along their length came to me at a woodturning seminar I attended two years ago, where I had a go on a bow lathe with the turning hooks as used by turners of old (see facing page). Also, David Ellsworth was there demonstrating the amazing cranked tools he uses to make his bowls, which are hollowed out through a small hole in the top (see his article, pp. 52-56). Woodcarvers have always used tools shaped like mine, and for the same reason—to get into places where a straight tool can't.

I have never liked scrapers. They blunt quickly and produce a rough surface. I use the conventional deep-fluted, U-section bowl gouge as a roughing-out tool. I finish with a shallow, long-and-strong gouge. It is ground straight across and used on its side, with the bevel rubbing for a slicing action. This means it can reach only so deep into the bowl before the bevel loses contact.

I forged the gouges shown below from straight long-and-strong ones. The forge was the woodstove in my living room; the fuel, barbecue charcoal; the draught, from a windy day. By blocking most of the grate with firebricks, I didn't have to use too much charcoal. For an anvil I set a blacksmiths' fuller in a lump of wood—a fuller is a rounded iron wedge that fits into the square hole in an anvil, for shaping iron. I heated the gouge red, placed its end over the fuller, and struck it with a heavy hammer. I had to repeat the process several times to get the full curve. To reharden the gouge, I heated the newly bent end bright red and plunged it into a bucket of cold water. It's important here to hold only the cutting end of the tool in the hottest part of the fire; otherwise thin tools tend to warp and thick ones may crack. Then I ground the tool smooth with small grindstones in an electric drill. I tempered it by heating it in my kitchen oven for one hour at 350°F, which brought the polished steel to a medium straw color, and then letting it cool slowly.

I always test new tools with a sharp masonry nail. If I can scratch the tool just behind the edge, I reharden and retemper. I have never found a chisel or a gouge that cannot be made excellent this way.

The tangs I either cut off short or bury deep in the handle. If the handle is securely joined to the wide part of the tool shank, turning tools cut with less vibration. For handles I like a simple cylinder, though I'll rummage among the tools I'm not currently using for a handle before I'll make one. I frequently make a new handle when the first one does not seem quite comfortable. For ferrules I have used short pieces of plastic pipe, but I often don't bother. The handle can still split when driven on with a ferrule in position. Metal ferrules are cold on the hands. A handle should not be too long, too short, too fat, too thin, too light or too heavy. These are all matters of personal taste.

I grind the bevels hollow, stopping just short of the edge. I hone, using a fine oilstone, several times before having to regrind. I round over the heel of the bevel, to prevent its rubbing and marring the wood, and to make it slide smoothly.

Recently I tried one of my bent gouges inside a large elm goblet. I was delighted with how quickly I could get it smooth, even though cutting from rim to bottom was going against the grain. These gouges also work happily a long way over the tool rest. No doubt dig-ins are possible, but using the gouges only as finishing tools, I've had nothing nasty with them. □

Douglas Owen makes his living as a turner near Bainbridge, England. Photos by the author.

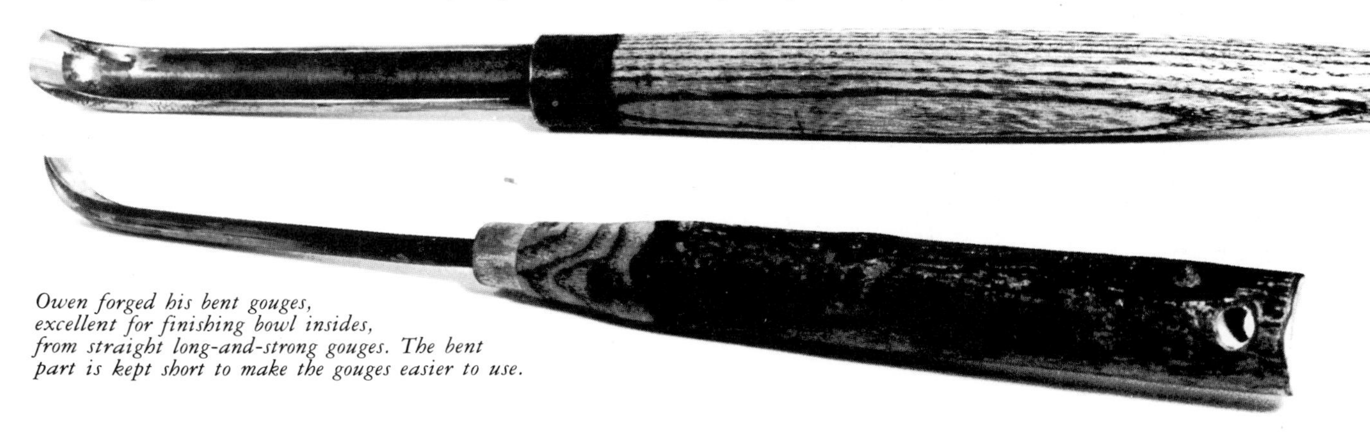

Owen forged his bent gouges, excellent for finishing bowl insides, from straight long-and-strong gouges. The bent part is kept short to make the gouges easier to use.

From *Fine Woodworking* magazine (May 1983) 40:95

Harvesting Burls

Strange formations are turners' delight

by Mark Lindquist

Kathy Lindquist

Bill Byers

Lindquist positions a gigantic yellow-birch burl in his woodlot for seasoning. Burls of this size (which can almost engulf a trunk, as on the ash tree in the inset photo) are difficult to move without the aid of professional loggers and special equipment.

Burls are perhaps the most misunderstood and mysterious material in the woodworker's realm. Most woodworkers and suppliers who use these strange, unpredictable organic formations have very little idea of how they grew or what caused them. And, although burlwood has long been prized for veneer, lumbermen and forest pathologists trying to produce straight and sound stock for the mass-production lumber industry have always regarded the burl as an enemy and a nuisance. But I find burls to be an extremely rich source of material for sculpture and turned and carved bowls, and, most importantly, an inspiration for discovering new philosophies and approaches to woodturning.

Burls are protuberances on a tree that come in all shapes and sizes. They are known colloquially as "burrs," "bumps," "knobs" or "gnarls," and scientifically as "galls." A burl may be a perfect half-sphere on the side of a tree, or look like a

From *Fine Woodworking* magazine (July 1984) 47:67-71

Kathy Lindquist

Bill Byers

Burls come in all shapes and sizes. Shown above are burls from 16 different species, including walnut, hornbeam, ash, beech, spruce, mulberry, birch, cherry and oak. Lindquist chainsaws bowl blanks from rough burls (left), visualizing the final form inside the burl. Notice the burl's layered structure. Below, the chaotic grain of a wild-cherry, dormant-bud burl struggles with a lathe-smooth surface. Linquist's father, Melvin, turned the piece, called 'Ancient Ceremonial Form Translation,' in 1974.

Robert Aude

wreath surrounding the tree, especially if it grows leaves on short stems. Burls may be irregular, twisted and malformed; surfaces may be smooth, or rough and fissured. A burl may grow halfway around the trunk, creating a half-moon that makes a beautiful carved bowl, or along one side of a tree, or sometimes all the way up and around the entire trunk.

Burl forms offer many sculptural possibilities. Some burls are quietly round, whether shallow or deep, definitely suggesting a bowl. Others are twisted and convoluted, evoking a sense of animation. No matter how much energy shows on the outside, though, it is no clue to the incredible explosion of energy inside—a swirling, frozen pot of marbleized color, texture and structure, creating patterns too complex to understand or predict. It is one of nature's amazing mysteries—the starry galaxy within is greater than the form containing it.

A burl is not a healing over of a broken-off branch, but an irregular growth. Burls are often caused when fire, frost or something hitting the tree injures the cambium, which is the growth layer near the bark. (Interestingly, when trees are repeatedly burned on one side, they form burls on the opposite side.) Fungal or bacterial irritation of the cambium also causes burl formations, particularly the more gnarly, rough-barked ones. Some scientists suggest that a mutation or a hereditary factor, combined with environmental conditions, is responsible for burl growth. This might explain why several trees of the same species growing in a particular area (which quite likely are related) may all grow burls.

Burls generally are divided into two categories: root burls and above-ground burls. Root burls (called "crown galls" by scientists) are caused by bacteria, which possibly entered damaged areas of the tree from infected soil. These formations tend to be softer than above-ground burls because they develop within a controlled atmosphere of soil and water. Usually a large, round ball will form at the base of the trunk, and the roots will grow out of the burl.

Root burls, like above-ground burls, may occur on any tree. They form on cherry, white birch and gray birch in the East, and are extremely common in California on redwoods and manzanita. Redwood root burls may grow to be 12 ft. in diameter, nearly too big to handle. Most of the redwood "burl tables" made in California are constructed of root burls with extraordinary grain and pattern.

On the East Coast, maple and cherry commonly exhibit above-ground burls. Next in prominence is the birch family, then oak, followed by elm, beech, ash, butternut, hornbeam and poplar. Soft-wood burls are rarer, and not as interesting in grain-pattern development, nor will they take as high a polish as the hard-wood burls. California redwood and buckeye burl are exceptions, since they have magnificent grain patterns, although they do not take a high polish.

Photo below: Robert Aude; photo at right: Paul Avis

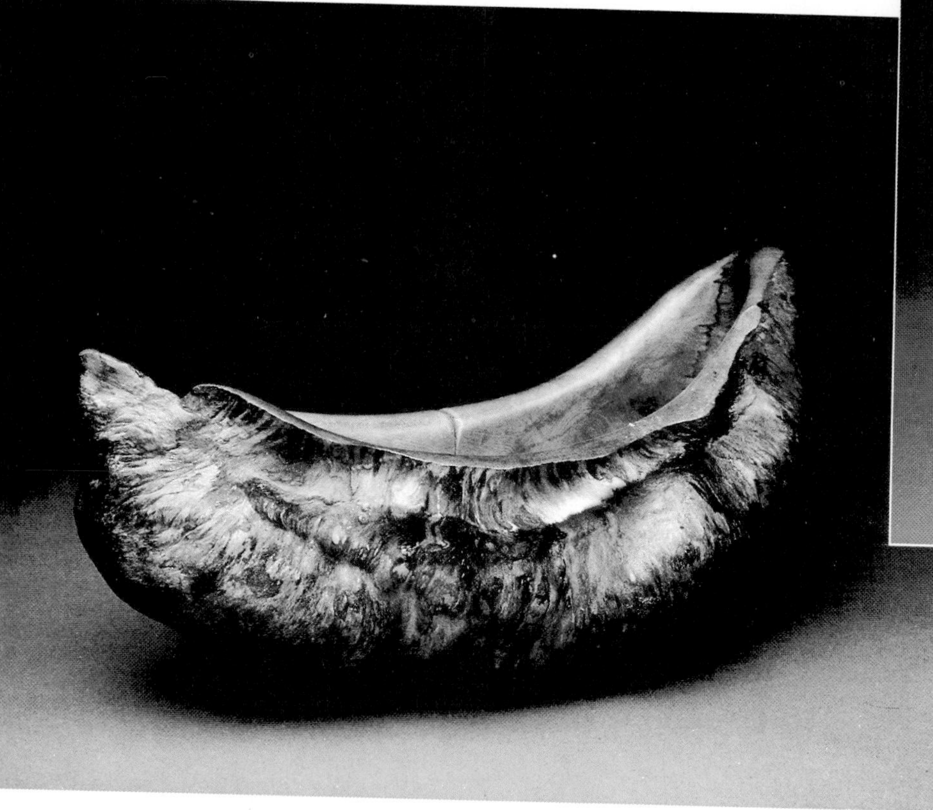

Lindquist worked the 12-in. long layered maple burl shown at left with a chainsaw, gouges, a die grinder, rifflers, rasps and foam-backed abrasive discs. Its oil finish is buffed with a tripoli/animal-fat-based compound. The piece, called 'Vessel Vessel/Arc Ark,' was made in 1978. A cherry burl yielded the 18-in. high turned bowl above. The burl contains layered and bud formations as well as bark.

An above-ground burl may grow at any height on a tree, on the trunk or on a major branch. Occasionally, more than one burl will grow on a tree, and when this occurs there are usually quite a few. The most exciting burl find is the "burl tree," a tree whose trunk has been entirely encompassed by burls growing into and around each other, forming a giant mass of burl growth and grain. Most burl trees I've found have been elm, maple or white birch.

Within the infinite variety of burl-grain development are three main classifications: annual layering, end-grain budding, and a combination of these two growth patterns. In annual layering, the burl grows in much the same pattern as the tree, though much more rapidly, adding a thicker layer than the tree, causing the bulging and swirling form of the burl. Successive layers within a burl will display inconsistencies in thickness, with thicker layers toward the center of the burl rather than nearer the supporting trunk or branch. Where the burl "hangs" from the tree, there is even apt to be a crotchwood formation in cell growth of the burl for support.

Compared to straight-grained wood, layered-growth burls display remarkable beauty and complexity of color and configuration, yet their grain is subtle compared to that of the spectacular burls formed by the end grain of dormant buds.

Dormant-bud burls form through an explosion of early bud development that never quite makes it through the bark. All the buds shoot and clash within, causing more shoots to get started and early ones to become dormant. The buds never get past the stage of early development, growing more in width than in length, causing hard, dense wood. This type of burl is most prevalent in cherry and walnut, and also occurs in certain elm, maple, oak and white birch species.

The third classification, which I call "swirl-eye" burl, is a combination of the other two. Often these are the most complex of all burl formations, especially when the patterns are balanced. Imagine that the first two classifications were melted together, then stirred and frozen. The dormant buds, or "eyes," are mixed in with the rest of the swirly grain.

After working with hundreds of burls during the past 15 years, I know that some grain patterns frequently recur within a given species. Yet on cutting into a new burl, I can still be surprised by an entirely new configuration.

Here are a few clues about where to look for burls. Although burls may grow on trees in any terrain, usually the hilly, rocky, heavily wooded areas of mountainous regions yield the most. An individual tree may give signs of burl formation—a gradual change in bark formation, for example, or a gentle widening of the trunk, indicating that a burl is growing on the other side. If you find a burl, keep looking nearby—burled trees sometimes proliferate in certain areas.

Perhaps the most difficult problem in obtaining burls is getting permission to take them. You usually can hunt down the owner of the land on which the trees you are interested in are growing and strike some sort of agreeable deal. Often I've traded small pieces of finished work for larger burls or lots of burls.

When I find a burl in the woods, I consider its quality, its distance from a usable road, cutting problems, and what I am willing to pay. Any woodworker who has harvested a big burl far from a road will probably not do it again, remembering the pain and difficulty of that first deep-woods encounter. One solution is to cut the burl in the fall and leave it in the woods until winter, then haul it out on a toboggan pulled by a snowmobile. Purists may take exception to cutting a burl or a tree during any time other than winter. I prefer the fall, as it is the best time to find burls because the trees are shedding their leaves—it's also the nicest time to be working in the woods. After years of searching for and harvesting big burls in this old-fashioned way, I've begun to buy them from local loggers, who frequently find them and are equipped to fell them and pull them out. Town dumps are often surprising sources, as are tree-removal services and tree surgeons.

A burl can be removed either by cutting it off the tree, or by cutting down the tree and then removing the section with the burl. If possible, I prefer the former because it eliminates dealing with the branches and other wood of a downed tree. Also, the tree normally lives on after the burl is removed, often becoming healthier. Don't paint the cut—it will dry up, and the tree will gradually grow around and over the wound.

A burl is best cut close to the trunk, as nearly parallel to the trunk as possible. Deep, gouging cuts into the tree are pointless when the best use of the wood is achieved by a clean, straight cut, leaving a neat area on the tree. You will always leave some good burled wood in the tree, but the only way to get it all is to cut down the tree.

When I do cut down a tree, I try to leave a foot or so of trunk on each end of the burl so that as the log dries, checks won't telegraph into the burl. Often the burl will dry better and is less apt to check if left this way for several years.

Drying and storing burls can be as uncomplicated as piling them outdoors or as sophisticated as packing them in layers of sawdust. When cut green, burls usually are very heavy, laden with water. Some woodturners prefer to use green burls. Burls turn most easily when wet, but will warp and twist as they dry. Green burls may be stored under water over a period of time, which prevents drying. Blanketing them in wet sawdust retards checking, promoting slow, even drying.

I prefer to work dry, aged, mellow burls—checks and all. My philosophy of woodworking acknowledges the nature of the material, which is to expand and contract. I allow for the inclusion of checking within my designs, by either accepting

Tips for turning irregular pieces

by Rude Osolnik

About thirty-five years ago, I started turning irregularly shaped blocks of wood because I had a good source of supply, and because of the challenge and the distorted, discolored grain. I've turned bowls, trays and weed pots from burls and spurs (where a tree root makes the transition from horizontal to vertical), as well as from found wood—chestnut fence rails, oak wagon tongues, just about anything.

I've found that a heavy lathe is crucial when turning irregular pieces. A large, off-center blank, 3 lb. to 15 lb. out of balance, will shake a light lathe apart before the piece can be trued. A heavy bed, preferably of cast iron, will absorb the vibration. Anchoring the lathe with bolts, sandbags or heavy metal bars will also help. (See Ed Moulthrop's lathes, pp. 46-51.)

I turn pieces up to 12 in. in diameter on an 800-lb. Oliver lathe. Even though the lathe is bolted to the floor, I sometimes wedge 2x4s between the bed and ceiling for extra support. For larger pieces, I use a 2,000-lb. metal-spinning lathe with a 13-in. swing over the bed.

Lathe speed is less important than the speed of the block at its periphery—the larger the diameter, the slower the speed should be. In general, I turn pieces up to 12 in. at 1200 RPM to 1800 RPM, and larger blanks at 700 RPM to 800 RPM. For blanks up to 24 in. in diameter, my big lathe turns as slowly as 150 RPM.

Before mounting an irregular turning blank on the lathe, you should trim, balance and shape the general form on the bandsaw or with a chainsaw. Bowls and trays are best mounted on faceplates. A weed pot or other turning where only the outside is shaped can be mounted between centers.

Roundnose chisel and gouge

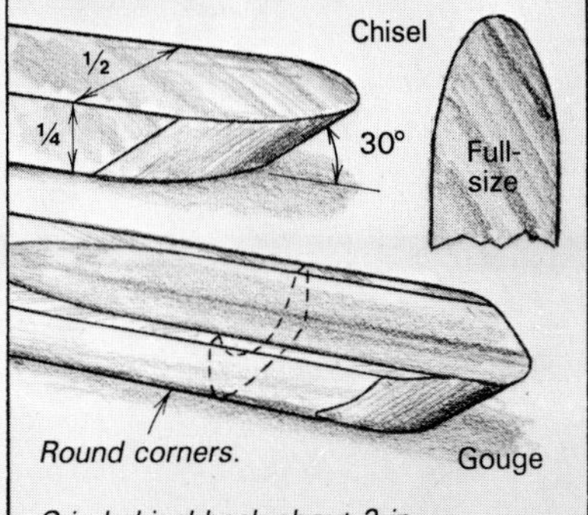

Grind chisel back about 2 in. from cutting edge to make gouge.

If there is enough waste material on the blank, or if the bottom won't be seen, screw the faceplate directly to it, using long, heavy screws. If you need a backing block, attach it directly to the blank with a strong glue such as Titebond. (Paper placed in between can separate.) Make the block thick—2 in. or more—so that you can work on the bottom without interference.

You can give a faceplate-mounted, deep bowl blank added support by embedding the tailstock center in the face of the blank. If the blank is very irregular even after you've trued the outside, you can leave a column of wood in the center attached to the tailstock while you shape the inside. After the bulk of the wood has been removed, cut off the column and finish the inside bottom surface.

I make all my turning tools, and I do most of my turning with a heavy ½-in. wide by ¼-in. deep roundnose chisel. Most people use this tool for scraping, but by grinding a long bevel on it, I can get the shearing cut usually associated with a skew chisel. The bevel, which is about 30°, rides the surface and supports the cutting edge, which can then take a smooth shaving. I also made a small, shallow gouge by grinding a groove in the top surface of a roundnose chisel, and I use it to take fine finishing cuts.

Truing up the odd shape of the rough blank is the hardest part. Once the blank has been trued and balanced on the lathe, it can be turned much like any other piece. Because of the irregular shape of the blank, the tool's cutting edge leaves and reenters the wood with every revolution of the lathe, increasing the chances of catching. Obviously, a steady tool rest and firm hand are essential.

Position the rest as close as possible to the blank. Be careful to stand to one side, out of the path of flying debris, particularly when turning burls. Remove the high spots with shallow cuts, moving the tool rest as you go to keep it close. Pay strict attention to where the edge of the work actually is. I watch the back edge where the stroboscopic effect shows the outline of the highest points most clearly.

As the tool makes more contact with the surface, you can take heavier cuts. For truing, I use a cut that is halfway between scraping and shearing. Posi-

the crack as it is, or working it open to relieve its jaggedness. I like the spontaneity of working the dry burl and being able to finish the piece, knowing it will move or shrink very little from its finished dimensions.

When I harvest burls, I usually leave them uncovered outdoors, sitting directly on the ground, for maximum exposure to the elements. Most begin staining and spalting, and I find the resulting colorations desirable. Yellow birch burls, for example, are relatively bland, but become utterly spectacular when colored or spalted.

It's a good idea to either mark the species and date of cutting on the burl, or store specific kinds of burls in separate piles so that they will all dry together. Painting of the sawn surfaces will discourage checking, but it also prolongs drying. I'd rather let the burls sit and let nature have its way. Since the grain is so twisty and gnarly, often the burls dry without checks, or with very few, which can be accepted, or cut out by dividing the burl along the cracks into smaller, usable sections.

After exposing the burls to the weather for a year or two, I move them into an open-air shed, build a lean-to over the pile using corrugated transparent roofing, or simply cover them with plastic. When the burls have sufficiently dried, they must be stored up off the ground, preferably away from sunlight. I pile them one on top of the other in my barn. Due to their various unusual shapes, this storage is awkward and cumbersome, as well as inefficient. I've built shelves and racks, but haven't solved the problem yet.

Like spalted wood, burls are abundant for those possessing the knowledge of their worth. These strange formations offer more than oddity or novelty for the woodturner, more than their hidden beauty. I find in them a profound truth, a truth about life and about the work of an artisan: in acknowledging imperfection, perfection is defined. □

Mark Lindquist turns and sculpts wood in Henniker, N.H. This article is adapted from his forthcoming book on sculpting wood, to be published by Davis Publications, Inc., 50 Portland St., Worcester, Mass. 01608.

Osolnik turns a rhododendron-burl weed pot between centers with a roundnose chisel (left). The nearly finished pot is shown above. The hole in the neck is bored after turning.

tioning the rest slightly above dead center, I angle the tool about 10° to 15° above horizontal. I don't worry about the condition of the surface until the blank is basically round. Then I resharpen the tool and take light, shearing cuts to get the final outside shape.

Normally I work very loosely, by feel, letting the grain pattern and shape of the block dictate the final shape, accenting areas that will enhance the finished piece. I like to leave bark on some pieces; the contrast of rough and polished textures adds character. I've even left moss in place on twig pots—the greenish color looks nice against the surrounding finished surface.

On the inside, I work from thick wood to thin—that is, from the center out—to minimize vibration in the wall. Turn the bowl wall $\frac{3}{16}$ in. to $\frac{1}{4}$ in. thick, then go back for the finishing touches. Check the wall thickness frequently so that you don't unintentionally cut through at an indentation.

When turning between centers, position the centers by eye and turn the blank by hand to check its balance and distance from the tool rest. After taking several initial cuts, you can reposition the centers to improve the balance of the piece, or to accent a particular grain pattern or texture.

I sand the piece on the lathe, by hand or with a disc sander. With quick-change discs and the lathe at 800 RPM, you can start with 60-grit paper—which removes torn fibers—and work through 80-, 100- and 150-grit, finishing by hand with 180-grit. Finer-grit, higher-speed sanding would burnish and polish instead of sanding. The disc will bridge gaps in the surface, so if you want to soften those, hand-sand with the lathe running slowly.

Friction and centrifugal force will dry $\frac{3}{16}$-in. to $\frac{1}{4}$-in. thick wet wood as you work. Start with a coarse paper—a fine grit will generate too much heat and cause checking. When you're done sanding, immediately bandsaw off the facing block so that the wood won't crack as it continues to dry. □

Rude Osolnik, retired chairman of industrial arts at Berea College, Ky., is a woodturner and turning teacher.

Photos: Rick Mastelli

Wenge, Africa
dark brown and black
11-in., $35

Exotic Woods

Observations of a master turner

by Bob Stocksdale

[EDITOR'S NOTE: Early in the summer of 1976 veteran woodturner Bob Stocksdale had an exhibit of some 120 bowls at Richard Kagan's gallery in Philadelphia. We were so taken by his extensive use of exotic woods that we asked him to tell us about some of them, as well as about how he works. The bowls speak for themselves. All 1976 prices.]

I have three lathes—two Delta 12-inchers and a homemade big one that is built of steel I-beams and swings 31 inches inside the headstock. One of the Deltas has the headstock blocked up 3 inches so I can turn up to an 18-inch diameter inside. I do 90 percent of my turning on it as I have an exhaust fan just back of it to solve the dust problem. All three lathes have jackshafts for better speed selection. They also have reversing switches to aid in sanding.

Almost all of the decorative bowls, trays and smaller salad bowls are started on a single center screw or 6-inch faceplate. I use several different methods to do the inside job, sometimes even the single center screw, but more usually, for footed bowls, the three-jaw geared chuck. Trays usually have a block glued on the bottom, with newsprint between for easy removal.

I do most of my turning with two gouges, 1-inch and 1/2-inch standard tools of the kind also used for spindle work. The corners are ground back a long way so the tip is really a half-oval shape. I use a shearing cut. I never use the deep, long-and-strong style of gouge, because I don't need all that metal, and there's very little strain on the tool. In fact, I'd like to get some gouges made of steel that is only 1/8-inch thick, the ones I have are about 3/16.

From *Fine Woodworking* magazine (Fall 1976) 4:28-32

Boxwood, Cambodia, white, 6-in., $40

Boxwood—Penberthy Lumber Co. supplies me with boxwood. It is one of the nicest woods to work because it cuts so cleanly and has a sheen from the tools before it is sanded. It reminds me of an eggshell and I have not found a finish for it that doesn't kill the beauty of the wood so I leave it bare, knowing that people with oily hands and peanuts will leave marks on it.

Silkwood, Australia, light brown, 11-1/2-in., $100

Silkwood—A wood collector in Australia sent me this piece of wood. It is the most lustrous wood I have had. It is in the maple family but is not very common. This piece is from near the stump. It works nicely but the highly figured area does not cut smoothly so my gouges have to be sharpened several times.

Canalete, Venezuela, brown, 8-1/2-in., $45

Canalete—This is a very oily wood of the cordia family and is quite common in Mexico. It goes by many different names. For some reason it is very hard to get. I got this piece from a wood collector. It is easy to work, does not clog up sandpaper, and has a strong, pleasant odor.

Tulipwood, Brazil, red, 6-1/2-in., $50

Tulipwood—Here is another wood used by the French in their old furniture for decorative bandings and veneers. It is another member of the rosewood family and not available in this country. It's easy to work but the logs are badly checked so a lot of repairs are necessary to get a good bowl.

Goncalo Alves, Brazil, reddish tan and deep brown, 11-1/4-in., $50

Goncalo Alves—This fine turning wood is another that has an unusual silky feel as it is sanded. Very easy to turn but many of the planks and boards twist and contort in the dry kiln and many surface checks show up. Recently I was offered a huge log that is in Le Havre, France. It weighs a couple of tons and would cost around 50 cents a pound. Too much to buy sight unseen.

Laurel, India, brown, black, 12-1/2-in., $37.50

Laurel—Most Indian Laurel that I have is not exciting enough to work but this is a dog board from a veneer company and it has almost a bee's wing pattern. So I had to make a tray from it even though it was only 5/8-inch thick when I got it.

Blackwood Acacia, California, light brown, 7-1/2-in., $75

Blackwood Acacia—This bowl is made of local acacia and did not cost me anything. It is a difficult wood to work because the sanding and tool marks are hard to remove. This shape is a hard one to do too. When I roughed it out the top of the bowl followed the curvature of the log, as it does now. I enjoy the final result because the wood has so much luster and depth.

Coralline, India, red, 11-1/2-in., $100

Coralline—I bought this log (15 inches in diameter, 17 feet long) on pure speculation. It grew in India and I selected it in London, but had never heard of the wood and the dealer was no help at all and charged extra because I would not take four logs. It is very difficult to work and takes a long time to sand as it is tough and stringy. The end result is worth the effort and after I had used 80% of the log I found it a very good dye wood—now I save all the shavings.

Ebony, Nigeria, black, 10-in., $200

Ebony—This exceptional piece of Nigerian ebony came to me from Penberthy Lumber Co. in Los Angeles. It is not easy to come by such a fine example of this wood as much of it has lots of flaws. The log was sort of diamond-shaped on cross section so I cut it in two with a big bandsaw and got two bowls out of each section. This one was near the center of the tree and had much nicer grain than the other. This ebony is not real hard and it turns and sands without problems. Any cracks that might be in the wood can be repaired with epoxy and lampblack and they will not show at all.

Kingwood, Brazil, purple, 5-in., $60

Kingwood—This is the only bowl in those photographed that is turned on end grain so the center of the log is in the bottom of the bowl. The logs of this wood are quite small and round so it lends itself nicely to this shape and style. This wood was used for inlay bandings on old French Provincial furniture.

Cocobolo, Nicaragua, orange, red, black, brown, cream, 10-in., $350

Cocobolo—This ranks among the top five pieces that I have made. It must be a freak piece of cocobolo because it does not change color like all the other cocobolo I have had. Most of it will change overnight and gradually darken until the grain patterns disappear. This piece did not change with two months' exposure near a window. I designed it to get a few touches of sapwood and as much as possible of the fantastic grain patterns that appeared just under the sapwood. An easy wood to work in spite of its hardness. Many people are allergic to it but luckily I am not one of them. □

Turning Giant Bowls

Ed Moulthrop's tools and techniques

by Dale Nish

When I first saw one of Ed Moulthrop's 36-in. diameter bowls, I was intimidated at the thought of turning it. It had evidently required a tree trunk for a blank, about a half-ton of green wood. I have since visited Moulthrop in his Atlanta shop, to learn about his monster lathes, harpoon-size tools and sophisticated techniques for controlling moisture-related wood movement in such hefty treen.

Moulthrop produces about 250 of these bowls each year, marketing them in craft galleries in New York, Atlanta, Scottsdale (Ariz.) and San Francisco. Although he's been turning since he was 14 years old, it was 10 years ago that he quit his thriving architecture practice of 30 years to do it full-time. His work has been exhibited in more than 40 art museums, including the Smithsonian's Renwick Gallery and the Vatican Museum, as well as being part of the permanent collections of New York's Museum of Modern Art and Metropolitan Museum of Art.

Moulthrop uses only southeastern woods, exceptional pieces of tulip magnolia (yellow-poplar, *Liriodendron tulipifera*), wild cherry, sweet gum, white pine, black walnut and orangewood, magnolia, and persimmon. He feels that native woods are amply exotic, if you find the right logs. His bowls are limited mainly to a few basic shapes: hollow globes, for which he is best known, lotus forms, and chunky donuts. The simplicity of his designs serves to enhance the elaborate, colorful figure of the wood.

Says Moulthrop, ***'Each bowl already exists in the tree trunk, and my job is simply to uncover it and take it out. I love the heft and the solidness of these huge blocks. I love to feel their weight as they resist the leverage of a big cant-hook, or to sense the tug of gravity as the hoist slowly separates a fifteen-hundred-pound block from the ground.'***

Moulthrop builds his lathes (he's built five in an improving series) specifically for large faceplate turning. For rough turning, he uses the one drawn on the facing page (and pictured on p. 49); for finish turning he uses a similar design, but of lighter structure (p. 51). In the former, the base is ¾-in. plywood glued to 2x4 supports in the corners. The top is a 24-in. by 35-in. section of a 1¾-in. thick exterior solid-core door reinforced with angle irons along its top edges. Moulthrop metal-turned the headstock from a scavenged, 3½-in. diameter steel shaft mounted in giant pillow blocks 18 in. apart. The centerline of the shaft is 38 in. from the floor, a comfortable height for Moulthrop, who's about 6 ft. tall. The tool rest, positioned a little below shaft center, is attached to a 5x7 solid cherry beam. The beam slides in and out from under the table, and is held in position by two large clamps.

Inside the base, a 2½-HP gear motor, controlled by a foot switch, is mounted on a hinged table, with the weight of the motor providing tension on a heavy ¾-in. wide V-belt. The motor output is 80 RPM. With four 9-in. pulleys on the motor, and four pulleys on the headstock shaft approximately 15 in., 10 in., 8 in. and 6 in. in diameter, speeds range from 50 RPM to 120 RPM. This may seem slow, but on a 30-in. diameter bowl, 80 RPM means a rim speed of 628 ft. per minute, about the same as a 6-in. bowl turning 400 RPM.

Moulthrop's tool rest is made from a 16-in. long piece of ½-in. by 3-in. by 4-in. angle iron, bandsawn to shape. A steel connector bar (⅞ in. by 2 in. by 16 in.) pivots from the cherry beam, cantilevering the tool rest in various positions. The top edge of the rest is drilled with a series of holes that take an 8d tempered nail, which serves as an adjustable stop to lever tools against, similar to the way tools are used on a metal-spinning lathe. A pin projecting from the bottom of the rest locates a support, which braces the rest against the floor when it's extended far from the beam.

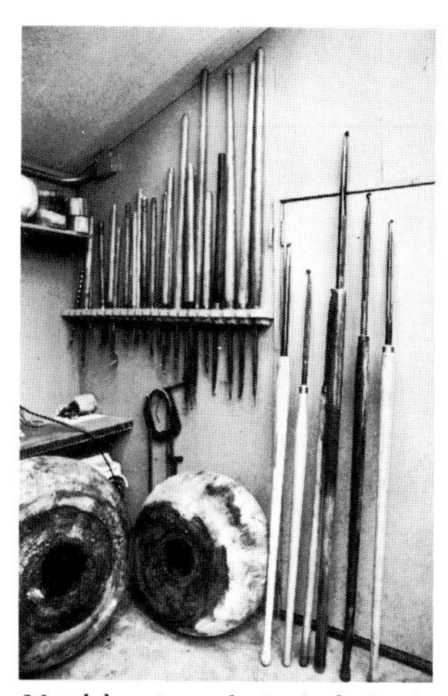

Moulthrop's tools include hooks and lances up to 96 in. long.

The three basic tools Moulthrop has designed and developed for his work are the lance, the loop and the cut-off tool. All are forged from either salvaged tapered-reamer stock or hex-bar tool steel,

---

Dale Nish teaches industrial education at Brigham Young University in Provo, Utah. He is the author of Creative Woodturning *and* Artistic Woodturning, *both published by the Brigham Young University Press, and is working on a book tentatively titled* Masters of Woodturning.

From *Fine Woodworking* magazine (July 1983) 41:48-53

Moulthrop's production includes bowls up to 40 in. in diameter.

Moulthrop's roughing-out lathe

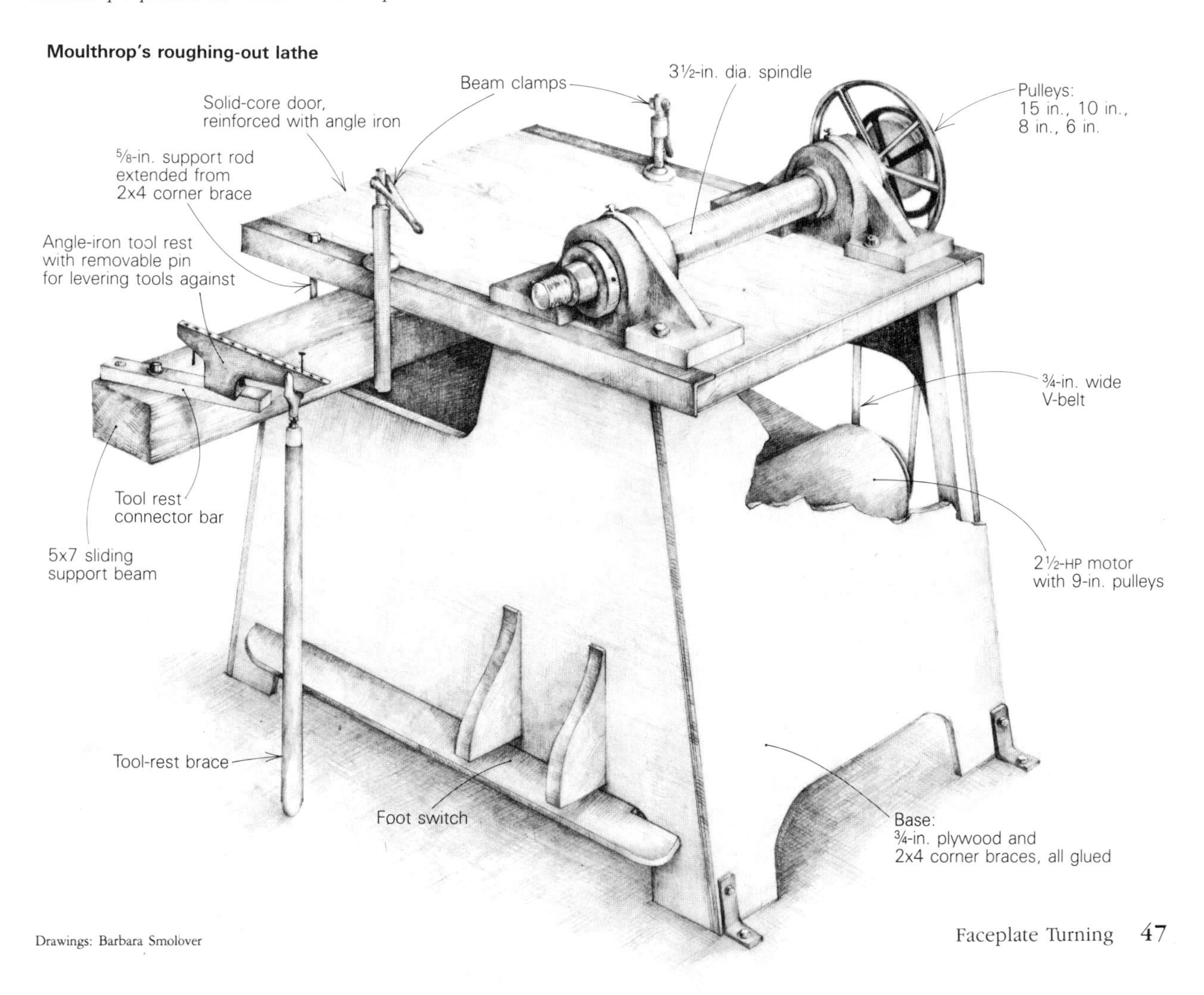

Drawings: Barbara Smolover

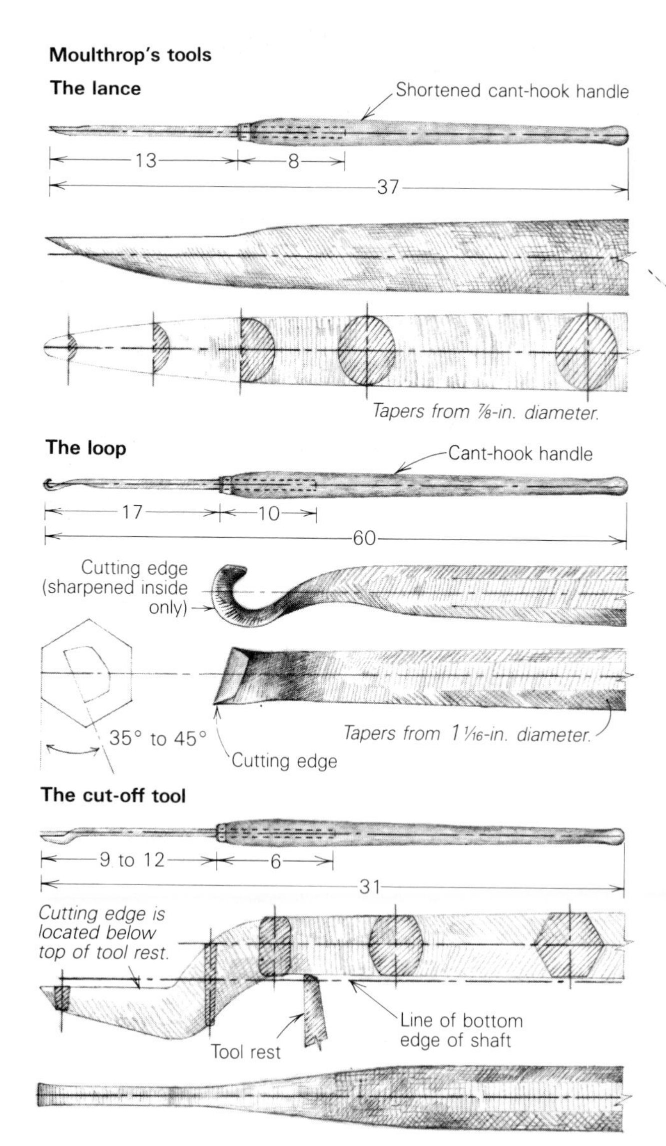

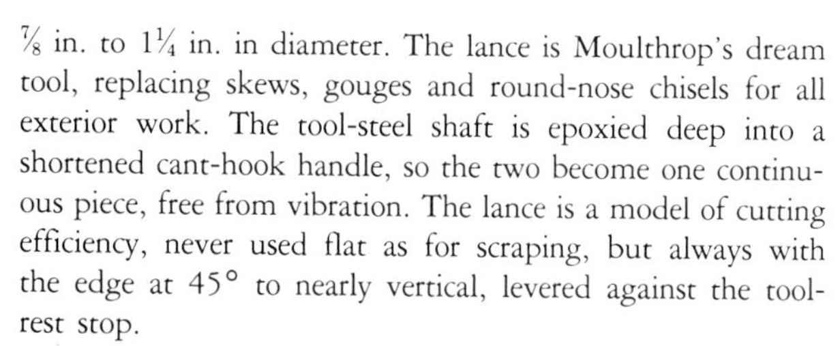

The wood yard, a grove filled with turning blanks.

Attaching the faceplate with lag bolts.

⅞ in. to 1¼ in. in diameter. The lance is Moulthrop's dream tool, replacing skews, gouges and round-nose chisels for all exterior work. The tool-steel shaft is epoxied deep into a shortened cant-hook handle, so the two become one continuous piece, free from vibration. The lance is a model of cutting efficiency, never used flat as for scraping, but always with the edge at 45° to nearly vertical, levered against the tool-rest stop.

For inside work, Moulthrop uses a loop not unlike the turning hooks used by turners of old (see pp. 33-34). After forging, the loop is tempered and then sharpened on the inside only, using a high-speed tool grinder and cone-shaped stone. Like the lance, it's used to cut, not scrape, always levered against the tool-rest stop.

Moulthrop's cut-off tool, a giant parting tool, is unique because its cutting edge is located below the level of the tool rest. The force of the turning workpiece thus keeps the tool aligned in the cut, safe from flipping over.

Moulthrop's wood storage yard is a grove of trees shading open areas covered by plastic sheets. Log sections waiting to be turned are left standing on end on the plastic, their weight forming depressions in which rainwater accumulates. Shavings heaped on top of the bolts and generously scattered over the plastic retain moisture in the rainy Atlanta climate, helping to forestall checking. The aim is to keep the bolts as wet as possible for as long as possible.

The damp climate also fosters staining and spalting of the wood. Colors seem to mature, sapwood darkens, and stains penetrate the bolt from both ends. All this adds an extra dimension of color and character to the pieces. Bolts may be turned fresh-cut, or kept for 6 to 24 months before they're rough-turned, still wet. In the interim, the older bolts may breed fungus, mold and even mushrooms. Before too long, however, these pieces will rot beyond usefulness.

Such large blanks require careful mounting. Moulthrop squares the ends of the bolt with an electric chainsaw, removing ½ in. of wood from each end so that he can see the color and figure. He lays out a circle on the end of the bolt delineating the best color and figure, not necessarily centered on the heart of the bolt. After chainsawing the bolt's diameter to its rough size, he rolls it into his shop, purposely located downhill from his storage area—a valuable feature, considering that some of the large bolts weigh 1500 lb.

The faceplate shown above was made from an old sprocket gear found in the salvage yard, Moulthrop's favorite shopping

First cuts with the lance braced on the thigh.

place. It is 9 in. in diameter and 3/4 in. thick. It has been threaded to fit the headstock shaft: 2-in. diameter, 8 threads per inch. The faceplate is carefully positioned in the center of the blank, and is mounted with heavy screws or 3/8-in. lag bolts, depending on the weight to be held. Moulthrop often drives additional lags between the teeth of the gear, as security against a massive mishap. After the bowl is finish-turned, the screw holes will be filled. Moulthrop uses epoxy mixed with wood dust to match the color of the surrounding wood.

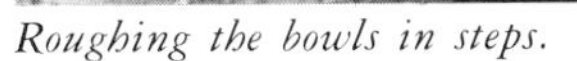

Roughing the bowls in steps.

In the typical roughing-out position (pictured above), Moulthrop holds the butt of the lance handle against his thigh so that the tool cuts at about the center of the piece. His right hand and leg steady the handle and control the tool's movement, while his left hand holds the tool firmly on the rest and against the tool-rest stop. His two arms and the tool form a triangle, providing maximum control as the lathe turns at 50 to 120 RPM.

The lance shears. The shank is held against the tool-rest stop and the rounded portion rides the surface of the turning work as the point penetrates the wood. After roughing the blank round, Moulthrop shapes the contour in a series of steps which are then faired. He typically removes shavings up

Smoothing the outside shape.

A. *Boring the clearance hole.*

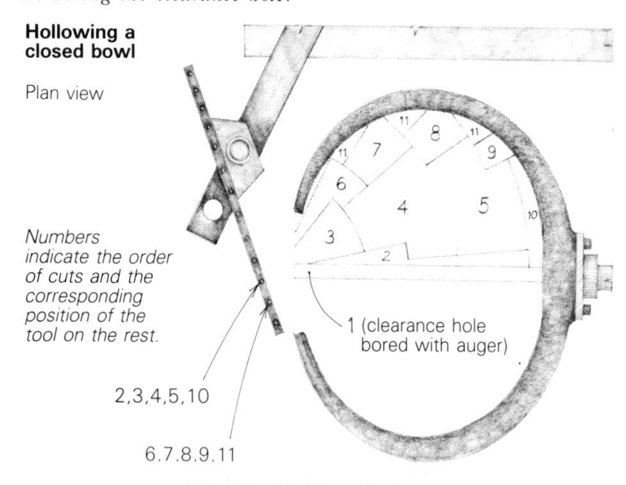

B. *The hook begins hollowing the bowl.*

C. *Undercutting the inside.*

D. *Several months in a solution of PEG stabilizes the wood.*

E. *After finish-turning (facing page), a flexible-shaft grinder smooths the outside. The dust pick-up cone, right, attaches to a 12-in. exhaust duct.*

F. *The flexible shaft also reaches the inside of the bowl.*

Using lance and loop, Moulthrop refines the shape of his bowls after PEG treatment, but on a lighter-duty lathe.

to 1½ in. wide and ½ in. thick. These peel off like long woolly caterpillars as water sprays from the moisture-laden log.

To hollow the bowl (facing page), Moulthrop first bores a 1¼-in. clearance hole with a brace and bit, held stationary while the lathe turns at slow speed. He marks the depth on the auger and drills to within 2 in. of the bottom.

The loop cuts from the center out, levering on the tool-rest stop, which must be repositioned periodically as the cut progresses. Wall thickness on small bowls is gauged by feel. On the larger bowls, where the reach is too long, Moulthrop drills ⅟₃₂-in. gauge holes and measures the thickness directly with fine wire. The rough bowls have a wall thickness of about ¾ in. at the opening to 1 in. or 1½ in. at the base. Later the base of the bowl will be thinned, from the inside, to ¾ in.

Polyethylene glycol 1000 (PEG) is the key to being able to turn such large bowls taken from anywhere in the log, even including the pith and knots, without the bowl splitting or warping. The PEG replaces the wood's moisture and chemically controls its dimensional instability. After green turning, Moulthrop dates the rough bowl and immerses it in a 40% solution, vats of which line his tank yard outside his shop. The 4-ft. diameter, 36-in. high aluminum vats are also from his favorite salvage yard. Soaking time varies with the temperature of the season: about 60 days in the summer (average temperature 80°), about 90 days in the fall, and about 120 days in the winter. Small bowls don't require as long a soak as large ones, but leaving them longer doesn't hurt. Because of the work invested in them, Moulthrop often allows more time for large bowls, just to be safe.

After soaking, the pieces are drained and set outside on a drying rack for a week in the sun. Final drying takes about two or three weeks, in a small room conditioned by a household dehumidifier.

The finishing lathe (above) is similar in design to the rough-turning lathe, except that it is lighter and has a smaller motor, since only light cuts are now made. The tools, however, are the same: the lance for outside work, the loop for inside.

Sanding begins with grits as coarse as 16 or 24. Both inside and outside are worked by hand or with discs on a flexible-shaft machine. Small holes and defects are usually patched with a mixture of sanding dust and hard-setting epoxy, just before final sanding with fine-grit paper.

The bowl's finish may be of several types; many are possible, but all finishes are sensitive to the moisture that PEG attracts. Moulthrop has been experimenting with his present finish for ten years now, but he feels it still needs further development. For those starting out with PEG, he recommends polyurethane as the easiest finish. It must be applied in conditions of low humidity. Moulthrop buffs with 0000 steel wool, followed by tripoli and rouge in oil applied with a cloth while the piece is rotating on the lathe. The last step is to remove the faceplate, patch the screw holes and hand-finish the bottom. □

Hollow Turnings

Bent tools and total concentration

by John David Ellsworth

Bocote (cordia) bottle is 8½ in. by 5½ in. and sells for $250.

Bowl turning is one of the oldest crafts. It is also among the least developed as a contemporary art form, compared to the advances in related media such as ceramics, baskets and glass. Mass-produced "Taiwan teak" salad sets at functional prices have flooded the public market, and the shop-class candy bowl still pleases Mother at Christmas time. Wooden bowls that don't hold oranges and apples are still a contradiction at many levels of modern society.

Conventional turning gouges and scrapers, however sophisticated, create obstacles for the bowl turner if his concerns are with the development of pure form, rather than with the juxtaposition of form and function. The key word is "development," pushing one's process and materials to a previously unattained limit, sometimes beyond. The bent tools and techniques I am presenting here may be unique to turning, but the resulting forms are quite familiar to artisans in other media, such as potters and basket makers. My bowls range in size from 4 in. to 16 in. across, and from about 1 in. to 12 in. high. Their walls are usually ⅛ in. thick, and sometimes I can go as thin as 1/32 in.—translucence. Success and failure are determined by the forms developed.

What do I mean by "form?" Every sculptor has his own relationship to the human body, and his own way of perceiving this relationship. The spontaneity of wood grain relates to the motion, tone, texture of skin; the cracks and decay to the imperfections of the human condition. I repeat forms many times, as different woods and their grains demand. With enclosed forms, my intention is to enhance the mystery of the interior. This allows the piece a sense of privacy within itself—a personal sensuality. The "function" of the piece becomes the simple interaction between observer and object.

Turning the outside — I work primarily with imported hardwoods and special cuts of domestic hardwoods—usually crotches, butts and burls. When I buy wood, I frequently select the garbage cuts as well as the select cuts. I can work green wood or dry wood, with long grain, cross grain, knots, sapwood, pith wood, whatever. My bent tools allow me latitude that I would not have with standard tools and methods.

I have a General 12-in. variable-speed (300 to 3,000 rpm) lathe, model 260-1. It is made in Canada and sold through C.A.E. Inc., P.O. Box 12261, N. Hwy. 73, Omaha, Nebr. 68112. It is comparable in price to Powermatic, but far superior in quality, built for business and with no racing stripes.

To mount the work on the lathe, I usually glue directly to a piece of ¾-in. plywood, then screw the ply to the faceplate with 1-in. No. 10 Phillips self-tapping screws. Heavier blocks are mounted directly on the faceplate with longer screws of the same type. When the piece is completed I go in with a parting tool and cut half plywood and half bowl stock, then break or split the bowl off with a chisel. This leaves a small spot on the bottom, which I remove with a 1-in. by 42-in. belt sander, an orbital sander and hand work. This leaves a surface with the same quality as the rest of the piece.

When roughing out a bowl, I turn at speeds from 300 to 600 rpm, but once the form is established, I jump to 1,800 to 2,000 rpm. All the designing is done on the lathe following the roughing-out stage. High speeds with a tiny-tipped, long-and-heavy roundnose tool allow me to draw on the surface of the bowl while removing great amounts of stock. It's spontaneous and leaves a clean, smooth surface that is ready to sand. Because I work with cracked and partly rotted wood, it is important to sand the exterior completely before beginning the interior cutting. I use silicon-carbide (wet-dry) paper because it breaks down evenly. Garnet paper would last longer, but it leaves minute scratches that seem impossible to remove. A grit sequence from 100 to 600 gives a beautiful surface in a very short time.

Cutting the interior — The beginning stages of internal turning are a simple process of clearing stock in an effort to get to the final surface. The principle is to enter with a standard angled skew and roundnose tools, open or clear the area with bent tools, and, once there is room to work, to finish the surface to the desired thickness.

My bent tools are made from square-section, oil-hardened ground stock tool steel, also called drill-core steel. I buy blanks 3 ft. long in four sizes (¼ in., 5/16 in., ⅜ in. and ½ in.)

From *Fine Woodworking* magazine (May 1979) 16:62-66

All prices below and in article are as of 1979.

David Ellsworth

Rotted figured walnut, 1½ in. by 10 in., $250.

Putumuju, 3 in. by 5 in., $150.

Gabon ebony, 5 in. by 9 in., $450.

African blackwood, 4 in. by 8½ in., $450.

African blackwood, 1½ in. by 6 in., $150.

Cocobolo rosewood with Brazilian rosewood lid, 3 in. by 12 in., $350.

Quilted maple, 3 in. by 9 in., $250.

Rare African pink ivory wood, 3 in. by 4 in.

Bocote (cordia), 3 in. by 5½ in., 1/32 in. thick, $150.

Tulipwood, 3 in. by 5 in., $190.

Macassar ebony, 2½ in. by 5½ in., $90.

from Teledyne Pittsburgh Tool Steel Co., 1535 Beaver Ave., Monaca, Pa. 15061; the brand name is Warpless Flat Stock. A nearby machinist forms the tools for me. We are presently experimenting with high-speed steel, which is harder but may turn out to be too brittle. The angle of the bend is about 40°, depending on the tool's function. I use three basic tip shapes in each of the four sizes, and each tip has three cutting areas, as shown in the drawings: the forming edge, cutting tip and trailing edge. This gives me about 36 distinct cuts with 12 basic tools. I use the 1-in. by 42-in. belt sander to sharpen—a burr edge for rough cuts and a smooth edge for finish cuts. The steep angle of the bevel provides more mass directly below the cutting edge, minimizing chatter when working great distances from the support of the tool rest.

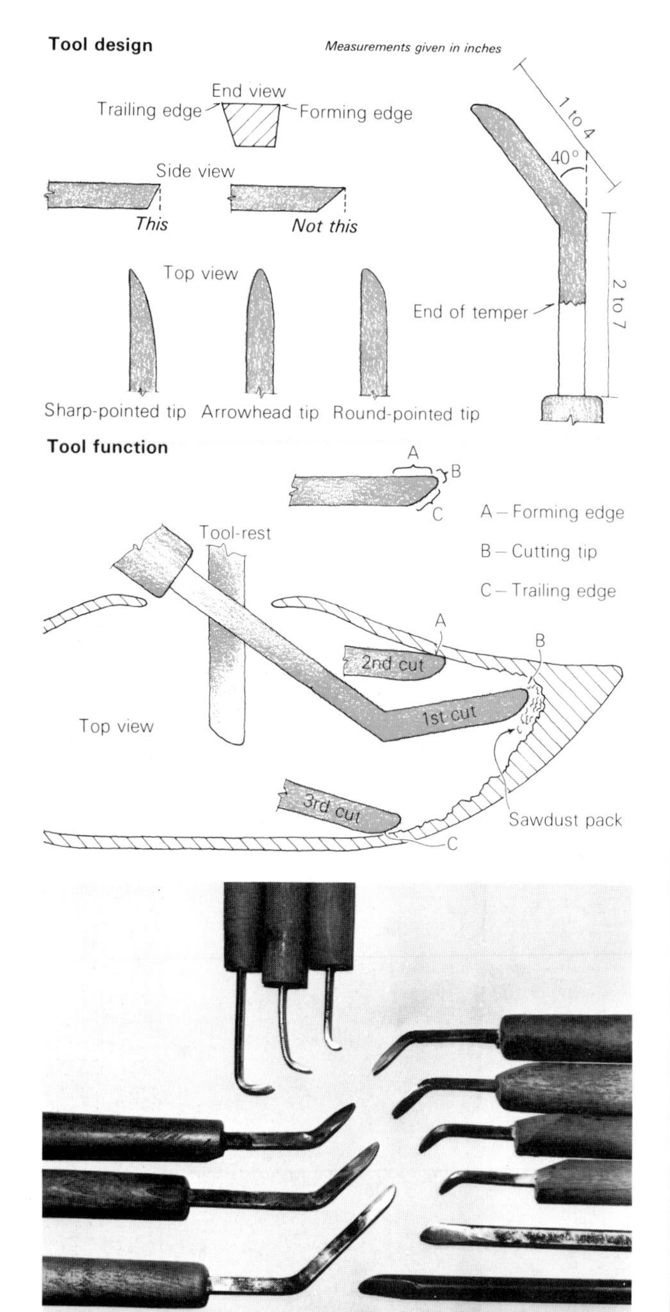

An assortment of bent tools. Ellsworth uses four sizes of tools and three basic tip shapes. The angle of the bend is about 40°.

When I start work on the inside, I keep the lathe speed around 1,800 or 2,000 rpm. Slower speeds drag the tool and increase the chance of ripping the wood. On rotten or cracked wood, the high speed prevents the tool from dipping into open areas before it re-encounters solid wall surface.

An efficient cut is a three-stage process, making use of each tool's three cutting areas. First, I use the tip of the tool to open up the area for working, which creates a rough surface. Second, I draw the forming edge of the tool toward the hole. It grazes the surface and removes the ridges left by the tip on the first cut, but it will not remove large amounts of stock. Third, I take a cut with the trailing edge from the thin wall toward the thicker area at the bottom of the bowl. Following this three-stage cut, I stop the lathe, remove the shavings and inspect the damage.

The straight-shank, pointed roundnose tool is very useful for reaching the bottom of deep bowls and for making final cuts across the bottom and up the sides. Because I am sometimes working as much as 12 in. from the tool rest, the tool must have length, weight and balance. The shaft is about 3 ft. long, made from ⅝-in. round steel shaft material, and ground as shown. All my tool handles are as simple and ugly as I can make them. The rough-turned, unsanded surface of red oak gives lots of traction for my hands. I simply split an oak 2x2, saw or rout out the halves to receive the tool shaft, reglue and turn down to desired diameter. Then I glue in the shaft with white glue, leaving 4 in. to 6 in. at the butt of the handle for weights to balance the tool, if necessary.

In tall pieces, the finish cuts are always completed before going deeper into the wood. This is important for two reasons: to eliminate mass at the point farthest from the support of the faceplate, and to use the mass in the lower portion to absorb the shock of tool contact against the upper portion. As I go deeper into the vase, I use heavier tools for increased support and control. I stop to

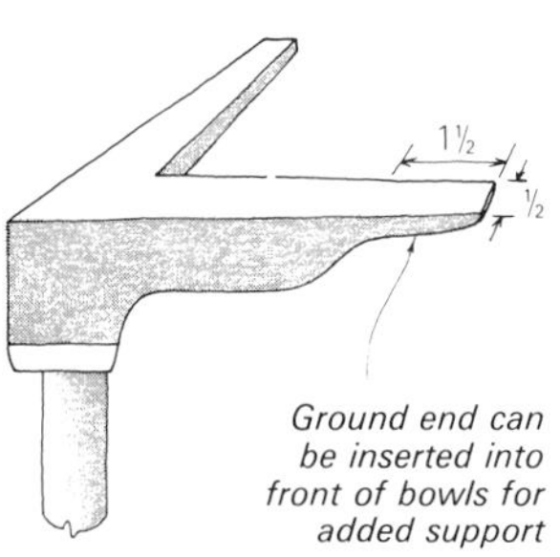

Ground end can be inserted into front of bowls for added support

Dr. J. Thomas Stocker

To shape the outside of a bowl, Ellsworth uses a 3 ft. long, ⅝ in. roundnose tool.

remove the shavings after each successful cut, to avoid tool drag. On open bowls, I loosen the sawdust with my fingers and vacuum or dump it out. On bowls with small openings, however, the sawdust pack must be loosened with a flexible wire, then shaken out. Whenever possible during hollowing, I insert the ground tool rest shown in the drawing on the previous page into the opening, for more support.

Determining wall thickness — My bowls can be turned so thin because an enclosed form is much stronger than an open form. It is like the strength of an eggshell. When turning at high speed with tiny-tipped tools, any single cut removes only a small amount of stock. A larger tool at a slower speed would only tear through the thin walls. I monitor the wall thickness as I work with a flexible wire caliper, and then by tapping the wood with my fingernail. The method is efficient and not difficult to learn. Even when I make a mistake and blow up a piece of wood, I no longer have to blow up *at* the piece, because I have fully known this period of its creation. Total control when working in the blind is neither possible nor desirable. Without the element of chance and the risk of error, the piece would lose some of its life—it would not have known this struggle for survival.

I start tapping at the edge of the hole, where the actual thickness can be measured with calipers, progress toward the rim and then down the sides to the bottom. A sharp tap with the fingernail raises a tone. When this tone is evenly balanced, I know the wall will be uniformly thin. I can then remove minute ridges on the inner surface by sweeping cuts with the trailing edge of the bent tools. The 3/16-in. steel rod I use for a flexible caliper is also useful for locating the actual area to be worked. In determining thickness it is only accurate to about 1/8 in., at which point tapping takes over.

The key to tool control is total concentration. This is why I rarely allow visitors into the studio when I'm taking a piece down to final wall thickness. I work each piece from start to finish by a series of controlled movements. These movements are gross in the roughing-out stage, but very fine in the thinning stage. It begins when I climb onto the machine, feeling its vibration and establishing a relationship of comfort and

Tapping

Measurements—side walls and bottom

3/16 flexible steel rod

Tap

1/8

Inside measure

Outside measure

Ruler

A – B = approximate side wall thickness

3/4 plywood

1 #10 Phillips self-tapping screws

Faceplate

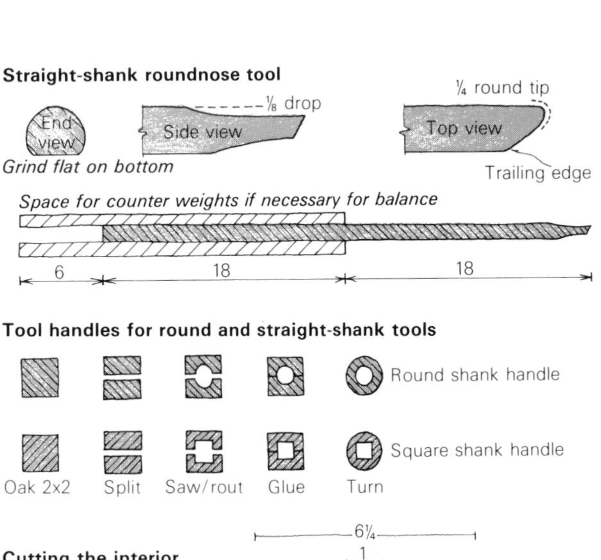

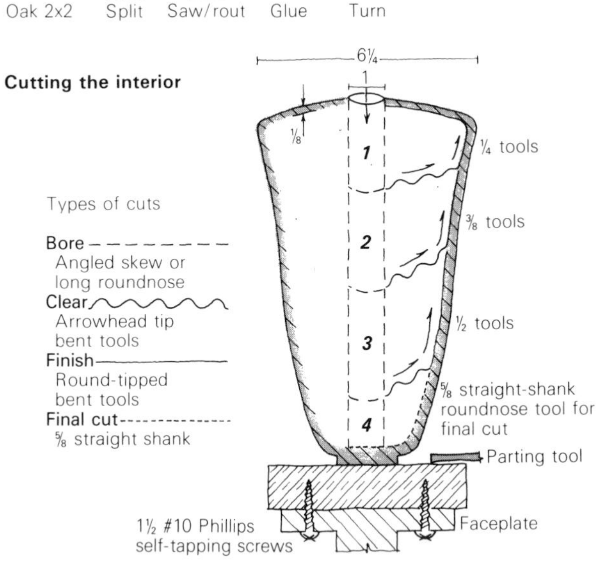

Total concentration is necessary for tool control. When Ellsworth straddles the lathe, each movement of the tool is a result of a movement in his entire body.

Internal turning begins with a ⅜-in. arrowhead-shaped tool.

The tool is supported on an angled tool rest.

Sanding

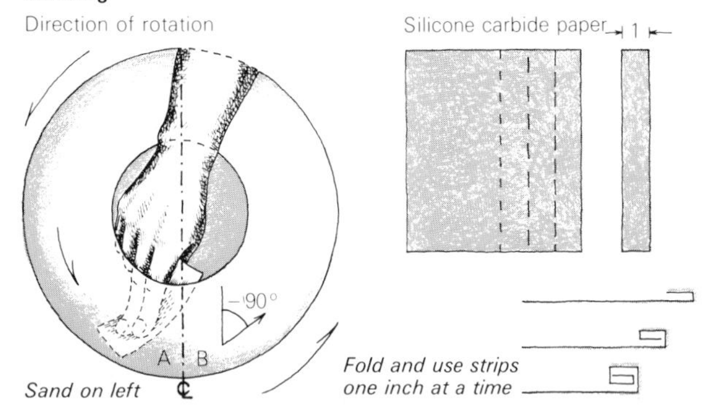

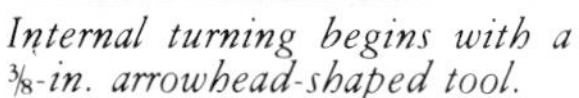

Cutting the interior of rotted or cracked wood

Each cut should be no deeper than ¼

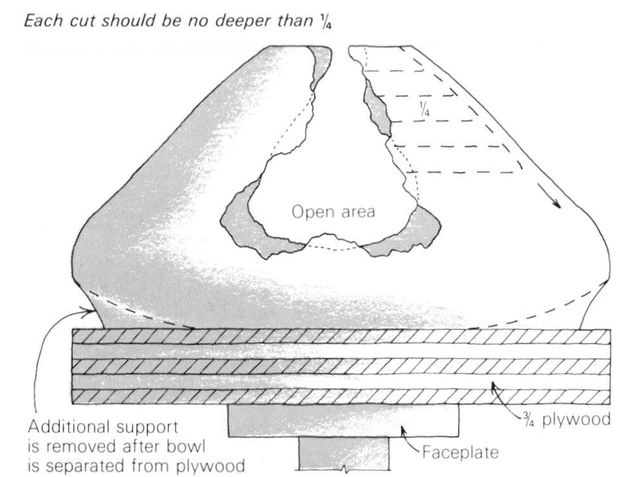

support as the turning progresses. Each movement of the tool tip results from a total movement in my body, not simply in the fingers, wrist or arm. I like to think of a cat stalking a bird: The concentration involves all senses equally, and the center of focus is transferred to the tip of the tool.

In the final stage of thinning the walls, tool contact with the whirling wood creates humming tones, and these become my clue to the consistency of wall thickness. The nature of the tone depends on the type of wood and on the mass remaining within the wall. With most woods, this tone becomes dull as the thickness approaches 1/32 in. Now the walls are very fragile at any single point, just like an eggshell. Any loss of concentration, including breathing at the wrong stage of the cut, can mean disaster. Thus sound, not accurate measuring devices, is the determining factor between success and failure. The only way I can describe it is that I talk to my pieces and they talk to me.

Rotten wood — The process of turning a rotted piece of wood is much the same as when working a solid block. More attention must be given to supporting the delicate walls, but I use the same tools and turning speeds. I always begin by gluing the blank to ¾-in. plywood. In the final stages of turning, this block gives extra support to the walls, especially if the rot extends into the bottom of the wood. I complete and sand the exterior before starting on the interior. I begin thinning the walls as soon as the entry hole is opened, leaving the bulk of the interior intact. I progress to the bottom of the piece, thinning the walls as I go, but never taking more than ¼ in. at a cut. If the tool seems to be ripping the wood or catching in the rot, the lathe speed is too slow. On the other hand, if I go too fast, centrifugal force could blow the piece apart. Experience is the only true guide.

Ironically, rotted wood offers several advantages. You can see the tool contacting the interior surface through the gaps in the whirling walls. The shavings fly out through the openings as the cuts are made, and wall thickness is easy to measure through the same openings.

Sanding the interior — I have a simple rule for interior sanding: If you can reach it, sand it. If not, learn how to leave very smooth tool marks. Sanding the interior is the only physically dangerous part of this process. I have broken two fingers learning to do it correctly. I fold strips of sandpaper as shown, folding 1-in. sections back against the strip as use demands. Support the strip between thumb and palm, allowing the area to be used to rest against the first two fingers. Insert the hand from the top, to make contact at the center of the bowl, and work the paper down and toward you, to the left of a vertical center line. This way the spin of the bowl is going away from contact with the sandpaper. If the hand drifts to the right, the angle of contact with the direction of spin falls below 90°. The surface is now coming toward you, and your fingers are supporting the sandpaper. One learns quickly.

I stop the machine between each grit and blow out the sawdust. Then I sand with the grain to remove circular scratches before moving to the next finer grit. The result is a beautiful interior surface, free of circular scratches.

Finishing — Once the final cuts are done on the inside, the piece is ready to finish. The following works well for kiln-dried woods, and it is essential for air-dried woods. I remove the bowl from the lathe, pour Penofin oil into the interior, work it around and pour it out. Penofin oil is made by the Penofin Oil Co., 819 J St., Sacramento, Calif. 95814. Don't use it on food containers. Then I remount the bowl on the lathe and with cotton sheet material coat the exterior with a liberal amount of oil. Crank the lathe up as fast as you dare, almost to where the piece blows up from centrifugal force. By working the rag all over the surface of the bowl I create a great amount of heat from surface friction. The hot oil begins to boil on the rag. Any moisture in the pores of the wood escapes in the form of steam. The boiling oil is then absorbed into the wood fibers previously occupied by moisture. As the rag finally dries, so does the wood and the oil within it. □

David Ellsworth, of Allenspark, Colo., taught sculpture until 1975, when he became a full-time bowl turner. He developed the methods described here after hours, when each day's production quota of traditional bowls was done.

Bolection Turning

How to inlay around a bowl

by Thomas J. Duffy

Bolection turning is the inlaying of a narrow wooden strip into a groove around a turned object. The word bolection means a fillet or molding with part of its section set into a groove and part proud. I extrapolated the notion from John Jacob Holtzappfel's *Hand or Simple Turning* (Dover Publications), whose process of inlaying a column economized on material and did not structurally interfere with the turning. Bolection turning has advantages over built-up work: Because the band does not support an adjacent section, it skirts the problem of different coefficients of expansion. The banding is not seen on the inside of the bowl, thus creating subtleties which can not be achieved in built-up work. And it is not difficult to do.

To begin, mount the bowl stock on a faceplate and rough-turn both the inside and outside to shape. Measure the outside diameter of the area where the band is to be located. For the band, mount a square of stock plankways on the faceplate. Once the stock is round, the band can be cut from the face and edge using a parting tool. The critical dimension on the band is its inside diameter, which should be about 3/16 in. less than the outside diameter of the bowl.

In order to remove the band in one piece, I've found that it is necessary to make the face cut first, and then tape the band area on the face to the material that will remain after the parting (1,2). I make the edge cut next. The stock that remains on the faceplate can be made into rings and saved for future work.

With plankways turning, the finished piece will have the quality of chatoyance, or changeable lustre and color. The end grain will pick up finishing materials differently than the cross grain. If uniform color and light reflection are desired, it would be better to cut the banding from stock mounted with the grain running parallel to the lathe bed.

Remount the faceplate carrying the bowl, and cut a groove the width of the band to a depth of 1/8 in. It is relatively easy to arrive at an accurate fit, but be careful to make it a "light feel" fit (3). A press fit swollen with glue will make installation of the band quite difficult.

Next, secure one half of the band on a workbench—use a bench-top clamp for large work and a vise for smaller work. Plankways rings must have the grain running parallel to the holding surface. The other half of the ring is held firmly at its axis and carefully snapped (4).

Bring the two sections together in the bowl groove (5). This should first be done dry. If the band fits, this should then be done with glue. Hold the band in place with a belt clamp positioned to allow a good view of the split areas, in order to be sure of proper reunion (6). After the glue has set, continue turning to finish off the bowl and shape the molding of the band. Take care not to break through the groove when dressing the interior.

If closed-grain wood is used throughout, finishing presents no particular difficulty. Problems arise when the main body of the bowl needs to be filled and the band shouldn't be (i.e., a mahogany bowl and a whitewood band). The dark filler will stain the whitewood, especially on the end grain. I haven't refined a method for diminishing this, but I've had pretty good luck carefully painting the band with the clear finish to be used and then proceeding with the filling. After the filler has set up, been wiped off and allowed to dry, the entire piece can be lightly sanded on the lathe, and finishing can continue in the desired manner. □

Thomas J. Duffy, of Ogdensburg, N.Y., is a self-employed cabinetmaker.

1 2 3 4 5 6

Photos: David B. Greenberger

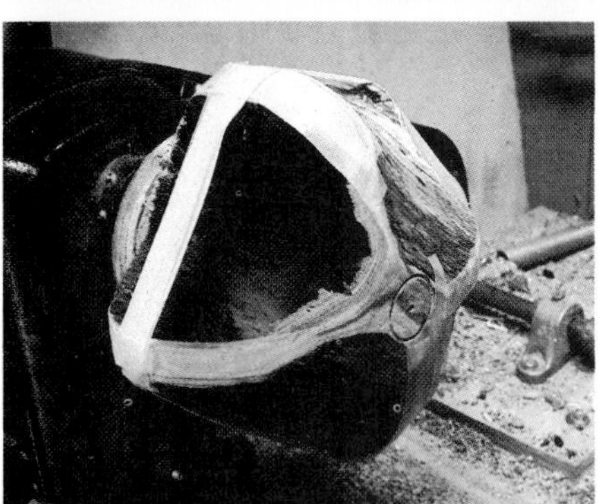

Turning a 4-poster

A 4-poster begins as a hollow log mounted on a faceplate (above). When the ends of the log have been roughed round (top right), they show the curves that will become the four posts as the inside is turned. To stabilize the wood for inside turning, Goff runs nylon filament tape up the posts and around the blank (left). Notice that the mouth opening has been scribed but not yet cut. The photo on p. 58 shows inside turning in progress, and the photo below shows the result. The wood is chinaberry.

Fig. 2: Turning a twistee

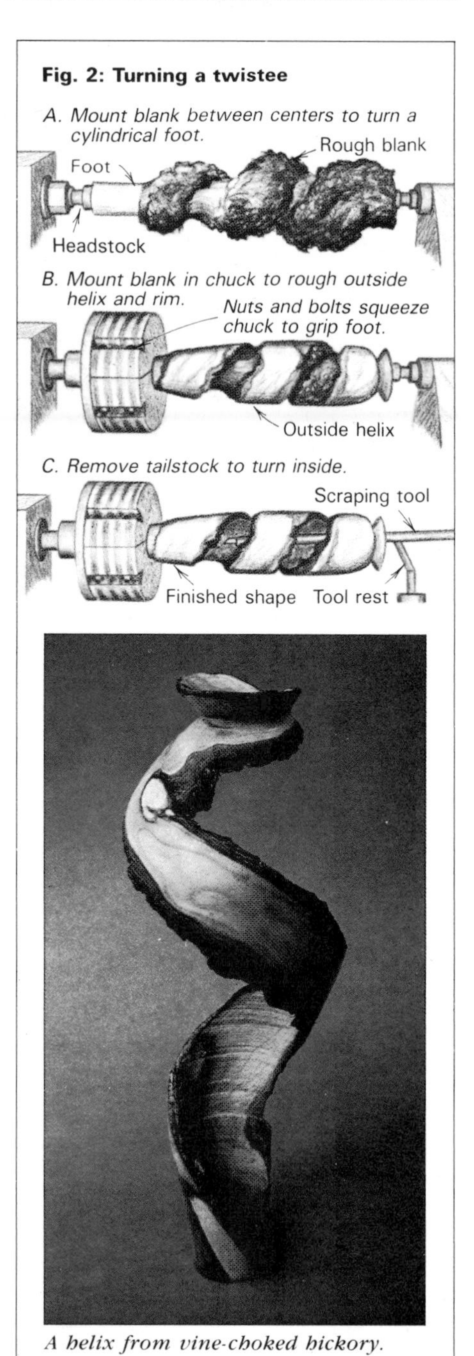

A helix from vine-choked hickory.

defect, but mostly it is just like the early stages of roughing down any irregular blank. As I gradually make the log round, I check that the four posts are symmetrical. If they're not, I shift the blank a little on the faceplate until things are working out right.

When it comes time to turn the inside, I reinforce doubtful wood with nylon filament tape—for a 4-poster I run strips up each post. Then I make a complete circle around the middle of the four posts. This causes the other three posts to add some support to the one being cut. I reposition the tape as the turning progresses. It's important not to distort the flexible posts when first applying the tape, or when repositioning it. Otherwise when you restart the lathe, the posts won't be in the same relative positions, and therefore won't be cut uniformly.

To start the inside, I either drill a hole or just cut through to the rotted inside with the cutting tools. The opening in the 4-poster on the facing page is 1⅜ in. in diameter. I then work down in shallow increments of ¼ in. to ½ in. at a time. As in other turnings, this allows the lower material to support the portion being cut. It's important to get the post thickness right before moving lower, as there will be no support if you try to move back up to touch up a thick spot. I find it necessary to turn the inside in one sitting. If the work is left overnight, even the least amount of warping will cause enough distortion to throw the posts out of line.

Another shape, which I call a twistee, is a helix. The lower right photo on the facing page shows one of the first I made. The stock isn't from rotted wood, but from a branch or small tree trunk with a vine wrapped around it. Over the years, the shape gets more and more pronounced until the tree finally grows back over the top of the vine. Finding trees with the most pronounced helical bulges is the first step in turning a twistee. I look for sections about 5 in. in diameter and 14 in. long.

I mount the wood between centers, as shown in figure 2, and reduce the headstock end to 2 in. in diameter, forming a round tenon that fits a shopmade plywood chuck. With the blank in the chuck, I support the other end with the tailstock so I can reduce the outside with the scraping tools until I have a clean band of wood spiraling from the base to the rim.

After the outside is done, the fun begins. I remove the tailstock and position the tool rest at the top of the vase. Then I turn through the top opening as before, but without the reinforcing tape—there's nothing to attach it to.

One problem is gauging the thickness. When I first began turning air, I liked being able to see the tool on the inside. But as I got to pieces with less and less wood, it became difficult to see exactly where the wood was. The effect is similar to watching a ceiling fan, where as the speed increases, the blades become a blur. Once—when turning the twistee shown in the photo—I had the uncanny experience of having the wood seem to disappear. I could see the top of the vase, with the tool shank entering it, and I could see the tool tip itself perfectly well, but the middle of the turning was completely gone. It seems that I had reached the point where 90% of the circumference was air; I *very* carefully finished it.

Sanding on both the 4-poster and twistee is done with the piece stationary. I begin with a grinder, then a drill. Next I hand-sand, using the sandpaper to work oil into the wood. This helps to seal the pores with the sawdust and to penetrate and toughen the wood, adding strength that is greatly appreciated by those who handle these delicate turnings. I have tried many oils, but the thickest and stickiest I've found—and the toughest when dry—is called Val-Oil, made by the Val-Spar Company in Stone Mountain, Georgia. The oil, usually four to six coats, is sanded in with wet-or-dry paper, from 100-grit up to 320-grit. The outside surface gains a fine luster, but I frequently leave the tool marks on the inside. Skeptics sometimes need to see them before they will believe that the piece was turned.

This 7-in. dia. birch bowl was turned from an asymmetric hollow log with one open side. The author frequently prowls the woods in search of such potential gems. The trick is to imagine what curious shape lies hidden in a rotten log.

A 4-poster bowl from spalted apple, about 11-in. dia.

All the wood I use is found, and part of the adventure is hunting through the woods with a hatchet, gently using it to probe the downed logs, looking for something that seems substantial. I have used about 35 different native southern species. Sometimes the logs are so far gone that it is difficult to tell exactly what wood I have in my hand.

Be forewarned that this sort of turning can have its unnerving aspects: Once, I stopped the lathe to reposition the tool rest, an operation I'd done five or six times already, and a scorpion jumped out and took off across the table. I was so stunned that I simply stood there and watched him scamper away. Scorpions, centipedes and black widow spiders all seem to like this sort of wood as much as I do. □

Gil Goff's shop is in Athens, Ga. Photos by the author.

Turning Spalted Wood

Sanders and grinders tame ghastly pecking

by Mark Lindquist

Spalted maple bowl (5 in. high, 11 in. dia.).

Just as spalted wood is not like ordinary wood, so turning spalted wood is not like turning ordinary wood. New methods must be found. Spalting refers to the often spectacular patterns of line and color that occur when water and fungus invade downed wood and begin to rot it (see more examples in the preceding article, pp. 58-61). These processes of decay change the wood's chemical and cellular structure, and its physical properties. In particular, the density of spalted wood varies enormously and unpredictably. Highly spalted zones are much softer than surrounding, more ordinary wood. The differences in density cause hideous pecking in turning, and spalted wood requires extremely tedious sanding before a smooth surface can be achieved.

My father and I, over the past decade, have developed an abrasive method of turning spalted wood. We rough out the blank with conventional turning tools, then, with the wood still whirling on the lathe, we shape and finish it with an assortment of abrasive body grinders and disc sanders. Our method also works on conventional woods and can be considered an alternative to traditional turning. But it transforms the working of spalted wood from an impossibility to a fascinating challenge.

I like to use pieces that have dried for at least three years after harvesting. Occasionally, if the wood was already dry when harvested and feels okay after a year, I attempt to work with it. But in turning the wood this soon, there is always a risk that it will crack or warp.

First, I use a Rockwell #653 power plane to expose the patterns and figures in the aged block of wood. It planes a swath 4 in. wide, has an optional carbide-tipped helical cutter (which I strongly recommend), and can handle just about any hard surface. I cut the ends off the block cross-grain to expose the end grain, which often is the key to the picture or figure of the spalting within. If the piece has been properly prepared and aged, cutting an inch or so off each end will usually take out end-grain surface checks and cracks. It is best to saw the blank so that the bowl lip faces the outside of the tree, and the pith side is toward the bottom of the bowl. This may seem extravagant, but it makes sense to have a bowl whose lip is less liable to crack later. I have also made reasonably successful bowls by turning end grain up towards the lip. The best way is to turn side grain, so that the bottom of the bowl has side grain and the sides or walls have end grain. Usually, the best figure or picture is in the end grain, so this is also an esthetic advantage.

Once you have visualized the bowl within the chunk of

Before turning, Lindquist exposes the pattern with a power plane, cuts out surface checks and cracks, and decides on the orientation of the bowl. Then the blank is bandsawn cylindrical and mounted directly on the faceplate with self-tapping screws. The bowl is turned between centers as much a possible and at low speed to avoid problems caused by the varying density of spalted wood.

From *Fine Woodworking* magazine (Summer 1978) 11:54-59

wood, examine the piece all over for checks and cracks. Determine which side is the bottom and which is the top by deciding whether the cracks may be turned out or eliminated With a compass, scribe a circle larger than the top of the bowl will be, and another that is the minimum size of the bowl. Somewhere in between, scribe the line on which you will cut the blank. Cut out the blank on a band saw—a ¼-in. skip-tooth blade works well—and center and mount the faceplate on the bottom side of the blank.

Normally, I use ¼-in. self-tapping sheet metal screws, ¾ in. long, to mount the faceplate directly onto the wood. This method has both advantages and disadvantages. There are usually harder and softer zones within a piece of spalted wood. While the wood is spinning rapidly on the lathe, it is easy for the tool to grab and catch in a soft spot that meets a hard spot. This wrench may be violent enough to rip the bowl loose from the faceplate. I've found that the more usual method, gluing the wood to a waste block with paper in between and screwing the waste to the faceplate, simply is not strong enough to withstand the wrench. In addition, the blank may be unbalanced because of its varying density, and screwing the faceplate directly to the wood is safer. Afterward, I plug the screw holes with turned pegs that over the years work themselves slightly out, but as they do so they also lift the bottom of the bowl just off the table, protecting the underside finish. Although gluing to paper isn't strong enough, a bowl can be glued directly to a waste block, which can be cut off after turning. And often the stock is large enough to turn a bowl with an extra-thick base, containing the screw holes, for cutting off later.

Once the faceplate is mounted on the bowl, and the bowl on the lathe, I chuck the headstock with the bowl up to the tailstock. Free faceplate turning is dangerous with spalted wood, again because of its varying density. Whenever possible, it is best to keep the bowl between two centers. Now with the lathe running at low speed, I rough the outside of the blank. I use carbide-tipped scraping tools, which I make myself; you can use conventional or carbide scraping tools or cutting tools. I prefer carbide because the black zone lines in spalted wood are very abrasive and quickly dull steels. Whatever tools you use, the first time you try spalted wood you will be shocked, disappointed and amazed, all at once. Spalted wood cuts like butter. It seems to be mush. But it pecks. Great chunks just tear out of the surface, leaving ugly pits and pockets behind. It will seem impossible to repair and you will want to throw the blank away. But don't, not yet.

Cone separation

First, rough out a fat version of the bowl you want to make. The inside may be removed by any traditional turning method, but with a large chunk of spalted wood the cone separation technique is a sensible choice. This trick removes an intact cone from the block of wood, which may be remounted and turned into another, smaller bowl. This is a good technique but requires confidence and care: If the tool catches deep inside a large blank, the lathe shaft will bend.

To separate the cone, I use a long, thick and strong file with its teeth ground smooth. I make the cutting edge on what was once the edge of the file, not the face, so the tool is rather thicker than it is wide. Grind the end to a short, sharp bevel, on the order of 60° or 70° and keep the corners sharp. Then add a long, sturdy handle for leverage.

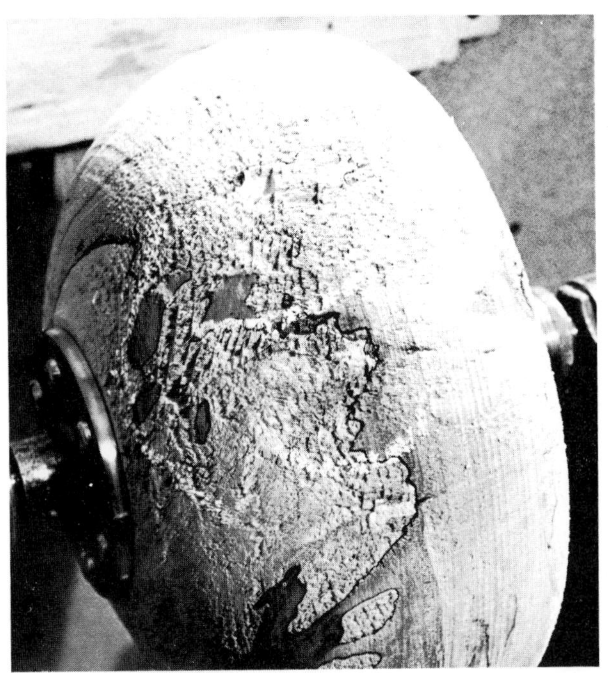

Spalted wood is a turner's nightmare. Rough turning yields pitted surface, with pockets nearly ¼ in. deep. Dark patches are hard wood; soft areas appear light. Black zone lines define patterns.

The lip and the face of the bowl are trued up with a long, straight borer, to prepare the blank for cone separation.

Start the cut at the inside rim of the bowl, and aim for the bottom center of the bowl. Make the cut at least twice as wide as the tool, for clearance. Once you have penetrated deeply enough to make a cone of the piece that normally would have been chips, use the same tool to begin cleaning out the sides of the bowl. The idea is to save the center section for another bowl or possibly two, and to keep the bowl between centers until it is almost completely roughed out. When the blank at last begins to look like a bowl inside (you can peer behind the cone, which is still attached at the center), take the whole thing off the lathe and place it on the floor, preferably on concrete. Using a spoon gouge and mallet, strike the center of the cone—with the cutting edge of the chisel facing end grain, not side grain—a good sharp whack. If you have done it right, a few subsequent blows will pop the cone right out. If you try to go at the side grain, there's a good chance you will split the bowl in half. Play with the method and find the best way for you—the point is to get the cone out without damaging the bottom of the bowl.

Once the bowl is remounted on the lathe, the inside must be cleaned up and the piece finish-scraped (not that it will do much good, for the more scraping that is done, the worse the

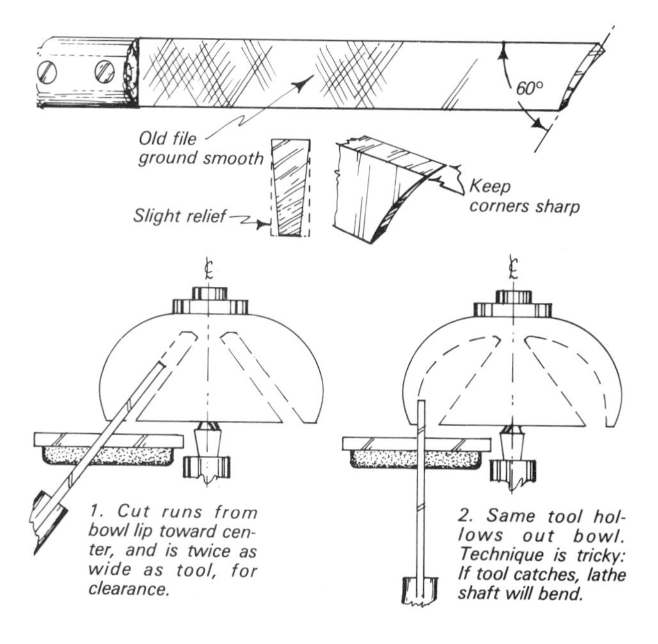

Cone separation: Above, a cut twice the thickness of the file divides the center cone from the bowl. After subsequent cuts have roughed out the interior of the bowl, the cone can be freed with a spoon gouge and mallet, below. A few sharp whacks of the mallet, with the cutting edge of the chisel facing end grain, will do the trick. The cone can later be made into another, smaller bowl.

chips keep coming out all over in what appears to be a turner's nightmare). At any rate, disregard the appearance of disaster, and let the bowl stand as it is, with all its blemishes.

Microwave drying

If occasionally I find that the bowl I've turned is still too wet to sand and finish, I leave it about an inch thicker than I want it to be all around. With white glue I paint the places that look as if they might crack—especially the end grain. If the wood is wet, the glue usually takes a while to dry and sometimes only becomes pasty. Then I pop the rough-turned bowl into a microwave oven. As I understand it, the microwave oven speeds up the molecules in the mass of the object, causing friction that results in heat. Not just heat, but a very even heat throughout the mass. Whole green pieces crack, but drier, merely wet pieces, especially in the rough shape of a bowl, dry out quite well without any cracking.

This approach is experimental, but here's the drying sequence I've found best: After I've turned the bowl over-thick and have it painted and placed in the oven, I try "shooting" it for 20 minutes on defrost. Defrost must be used, otherwise the bowl heats up too fast and is sure to crack. When the microwave is set on defrost, it cycles for a minute, then stops for a minute, then resumes, and continues this cycle until the timer stops. I've found that for bowls about 5 in. deep or 12 in. in diameter, 20 minutes is a good first cycle. After the first cycle, the bowl will heat quite rapidly and must be cooled. I usually leave it right in the oven for another 10 or 20 minutes, then check it for cracks, repaint it if necessary, then zap it again. Depending upon how wet or green the bowl is, it may be dry enough to finish after three or four cycles. Each bowl reacts differently, so times vary. If the piece is very special, I start out slowly (10 minutes on defrost), cool for another 10, then recycle all day. Usually the bowl will move considerably, which is to be expected, but cracks won't start.

I am still experimenting with microwave drying, and would like to hear from others who've tried it. Although the trick works in a pinch, I don't think there is any substitute for air and time. Thorough air-drying is the only way to allow all the richness and mellow color to come out.

Equilibrial abrasion

Torn end grain is always a problem in woodturning. In traditional turning, using as sharp a chisel as possible, gouging and shearing produce relatively smooth, "planed" surfaces. This requires much skill and a delicate balance between a sharp edge and perfect technique. My alternative uses industrial abrasive tools and products.

Turning hard, brittle spalted wood creates and leaves deep cavities and pecks in the surface. No amount of scraping with the sharpest of tools will improve the situation. When we first worked with spalted wood, my father and I would spend hour after knuckle-bending hour sanding through the pecks and chips to achieve a uniform surface. One spalted bowl would consume several sheets of coarse-grit paper and a lot of grueling work. Eventually, all the pecks and chips would disappear and the bowl could be finished. By the time we arrived at that point, all of the creative energy of making a beautiful bowl was gone, and it had become just a chore to endure.

After several years of searching and talking to other turners who were experimenting in the area, we arrived at a theory that we now call equilibrial abrasion. The shaft of a lathe

About bowls

I turn my bowls for appearance and artistic expression more than for utilitarian function. This may be a controversial approach among woodworkers, although it is in accord with artists and sculptors who accept a work for itself and not for its utility. As I see it, the bowl's function is to command the space of a room, to light its environment. Its function is to display the beauty of nature and to reflect the harmony of man. It is wrong to ask the spalted bowl to function as a workhorse as well, to hold potato chips or salad or to store trivialities. The bowl is already full. It contains itself and the space between its walls. The bowl is simply a vehicle in which the grain and patterns of the wood may be displayed. The patterns and colors are natural paintings, the bowl a three-dimensional canvas.

Complete and utter simplicity is required in the making of the spalted bowl. The simpler the form, the more uncluttered the surface for the wood to display itself. If we make bowls with lots of curves and decorative lines, the forms within the form fight with each other and with the wood. Wood that is spalted has become graphically oriented. To understand the art of making a spalted bowl, first understand the art of the ancient vessel. Study ancient Chinese and Japanese pottery vases, bowls, and tea-ceremony cups. Look at the work of Rosanjen. Investigate Tamba pottery. Study the masters and see simplicity at its very best.

—M.L.

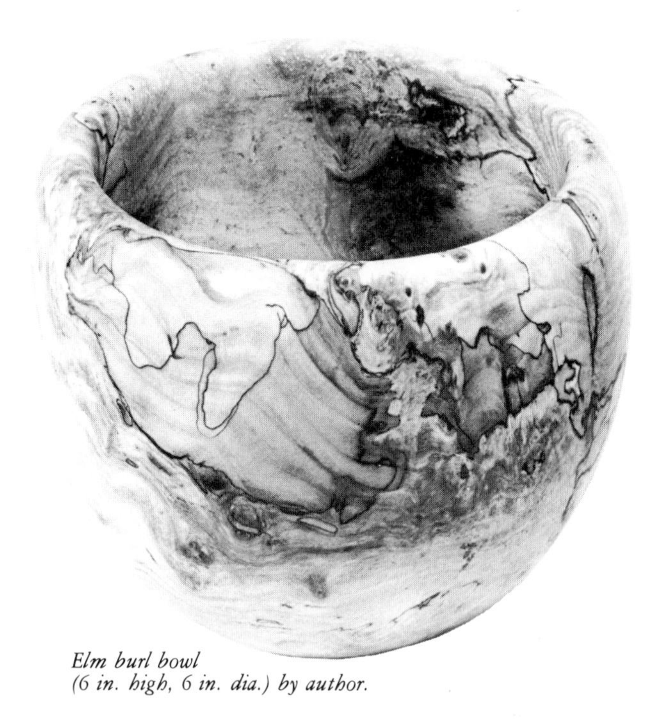

Elm burl bowl
(6 in. high, 6 in. dia.) by author.

turns clockwise. Most common rotating shop tools also turn clockwise. But if the shafts of two clockwise rotating tools are put together face to face, they oppose and rotate in opposite directions. We've found that two counter-rotating forces, with proper control and balance, can reach a state of equilibrium. We use abrasive discs on auto body grinders, rubber disc sanders and foam-pad disc sanders while the lathe is turning. Rather than holding the sandpaper still while the bowl turns, the spinning sandpaper works in the opposite direction against the spinning wood. The physics of the interaction aren't clear, but in practice there is a point—and it's not difficult to find—where the two rotational forces balance each other. The tool seems to hover over the work, and the sanding dust pours away in a steady stream. The counter-rotational system easily overcomes the problem of end-grain tearing. It does not eliminate the problem, it merely deals efficiently with it. And while this method won't replace the turning gouge and scraper, it incorporates the past and offers a new alternative for difficult woods.

Our method began in experiments with "flap-wheel" abrasives, at a time when it was difficult even to get flap wheels because the manufacturer, Merit Abrasive Products, Inc., was offering them only as an industrial product. Today they are also marketed for the hobbyist. We began by grinding the interior surface of a bowl—the biggest and most time-consuming problem—with a coarse-grit, 6-in. flap wheel driven by an electric drill. The outside surface wasn't so bad because it was accessible and conformed naturally to sanding. So, with counter-rotational abrasion using the flap-wheel accessory, the inner surface of the bowl was sanded much more easily than by hand. Soon I tried my auto body grinder with a coarse disc on the outer surface of the spinning bowl. At first, this was scary and dangerous, but after practice and with faith in myself and my tools, I found out how to achieve a perfect abraded surface. The tool and the application of it to wood were not new—but using the body grinder with the bowl spinning in the opposite direction and actually shaping the bowl with the grinder were new to us. With practice, the body grinder, a heavy and unwieldy machine, can become a sensitive instrument that improves with constant playing.

With such rapid sanding, the whole room immediately becomes full of dust, thick enough to cut with a knife. So an essential accessory is an efficient blower system located above and to the left of the bowl, facing it from the front of the lathe bed. Positioning is a matter of preference, but the best position is the one that sucks the most dust and is out of the way. The dust must be sucked up before it can enter the room, or it is impossible to work, breathe or see.

A simple squirrel-cage blower obtained from a local scrap yard will work fine for moving the dust. The best kind is a blower with a cast-iron housing, ball bearings, and at least a 6-in. diameter intake opening. Six-inch galvanized stove pipe works well and is cheap. The motor for the blower should be at least ½ hp and turning at 3,500 rpm. The on-off switch should be located near the lathe, since the blower is frequently turned off and on. An alternative to the blower is a large fan in the wall in front of the lathe to suck the dust out of the room. But I prefer the blower because I can direct the intake pipe to the area releasing the most dust.

After success with the body grinder, I got a smaller hand-held grinder that is easier to control, to use with finer grits. I use it with 120-grit floor-sanding discs to prepare the outside of bowls. I also use a small, flexible, rubber-backed disc on the interior of the bowl. Both Merit and Standard Abrasives make a system called quick-lock or soc-at that connects the sanding disc to the pad with a screw or snap fastener, making grit changes quick and easy. My usual grit sequence begins with 24, then 36 or 50, then 80, then 120 followed by finish-

Fig. 6: Bowl, Australian black walnut, 10 in. long, 1979.

Fig. 7: Chair, western red cedar, 34 in. high, 1974.

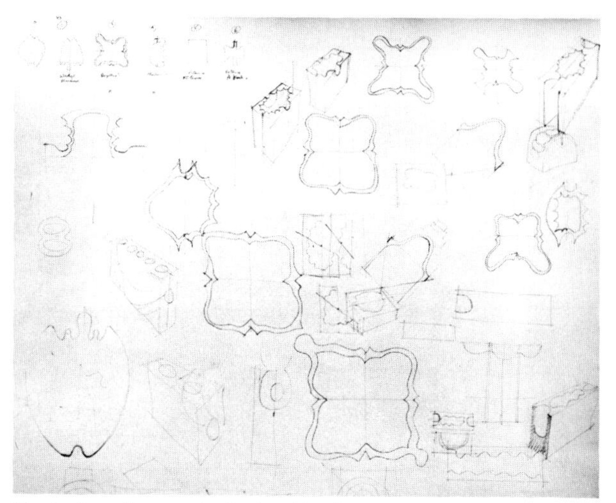

Fig. 8: Page from author's sketchbook.

another point of departure. This diagonal cut or wedge shape removed from the inside of the bowl made it asymmetric. This cut was internal. The next step was to use the cut to develop forms that leaned into, or projected from, the base of the piece; or, conversely, that presented the diagonal cut on their top surfaces. The piece illustrated (figure 6), which began as a square of wood on the faceplate, turned both sides, represents another idea made just last year, thus separating the first idea and this one by nine years.

As suggested by these examples, the system imposed by a diagrammatic analysis does not close other avenues. An apparently rational equation can be transcended by an intuitive leap into a fresh perception. The chair (figure 7) was made from a 6-ft. disc, flat on one side, turned on the other, with a hole in its middle. This wrinkled toroid was cut into quarters and reglued flat-to-flat, making blanks for two chairs. The cuts for the seat and back released two blocks of wood that I made into a small table; the scrap I had to remove to make the tabletop horizontal left me with two elements that suggested shelves. The sequence required quartering the circle plus additional diagonal cuts to unfold the suite—two chairs, a table and two shelves from a single large turning. This kind of complexity can easily obscure the original turned form.

Clearly developed forms that do not conceal their technical origins seem to have a sense to them. When the form also carries symbolic meaning, then process and form both become subservient to another cultural force. The sense or expression of any object may feature its structure, construction, material, shape, function, purpose or meaning, with each aspect responding to economic considerations as well. By "economic" I mean the expedient solution to a design problem, compared to the extravagant solution—understated or overstated, understructured to overstructured. An example might be a shooting stick compared to a pilot's chair. The former is an expedient solution to the problem of sitting, with minimal comfort; the latter is extravagant, maximum comfort. A design that works is an optimum solution to some particular set of circumstances and requirements.

All objects, however, record the motivations of the maker, and motivation can also be understood (interpreted) in terms of the criteria listed above. Usually one aspect of motivation is dominant, and this becomes the expression of the designer's intent. All design is a compromise—I am clear about the disadvantages within my chair design (for example, its weight), but I endeavor to understand its primary purpose: the application of forms of this kind to architectural spaces.

Along with this study of geometric possibilities, I inquired into the cross sections of these segmented turnings—that is, the ribbon of shape surrounding their volume, like the clear meat surrounding the heart of fruits and vegetables. The skin, shell or outer casing of organic material is often simple, consistent and even, with the visual pattern developing toward the center. The crimped edges of shells, the wrinkled edge of nuts, and the fluid lines of oil on water suggest a dynamic relationship to the regular curve of the circle. These articulated edges of energetic growth add mystery to the totality of the circle (figure 3).

Ideas seem to present themselves in three basic ways: through seeing, drawing and imagining. We all look but how often do we see? Making connections between two apparently unrelated situations is "seeing" in the way I use it here. For one table design I remember seeing a bird taking off; it appeared unmistakably as a turned form. The rapid motion of the wings appeared to be the outline of a rotating disc (the table is a sectioned turning). Drawing and sketching evolve shapes and relationships on paper that may be too complex to juggle mentally. Imagining is discovering relationships within one's own mind. Designers use all three approaches to idea generation, and switch comfortably from one to another. I find that whatever method seems to be working best at the time, I still end up drawing my ideas before I actually make them (figure 8). Drawing also has the advantage of freezing an idea for extended scrutiny, and it is a quick way for a woodworker to test a form without investing wood and work. It also develops sensitivity to the proportions, feeling and sense that a design will carry.

To conclude, I would like to introduce my concern for the purpose of making forms like these. As investigations of technique suggest ways of making new forms, the questions of why this or that shape, and to what use in daily living these forms can be put, become paramount. There is not space here to debate the whole hierarchy of economics, structure, construction, material, form, function, purpose and meaning.

Wood is an amazing material, but metal and plastics also have properties that make them superior for some uses. While I work primarily in wood, I do not like to see it wantonly misused in inappropriate ways. Some of my designs (e.g., figure 5) were intended to contain soil for plants, perhaps outdoors. Concrete or aluminum would be more suitable than wood, but I could not turn them on my lathe. I therefore turned a blank of expanded polystyrene, then had a foundry cast it in aluminum using the burn-out method, similar to the lost-wax technique of the jeweler. The concrete forms were never cast, even though I did develop a plug that would generate a bicycle rack. Lack of knowledge, time and resources stopped that development at an early stage.

I have also made a number of small sculptures in limited editions. First I turned the form in wood, then had it translated into aluminum by sand-casting, and then into polyester resin by casting into a mold prepared from the wood blank. It is very useful to see the same form in different materials, colors, textures, weights, etc., as the form can thus be studied independently from the seductive nature of wood. It is too easy to rest on the naturally beautiful qualities of wood.

Another aspect of my inquiry was the relationship of these forms to architecture. Contemporary public architecture is often technically superb and beautifully proportioned, yet lacking the human qualities of texture, diversity and the responsive hand and spirit of the craftsman. Rather than agitate for a total change in building technology, I advocate adding an overlay of decorative elements that display texture and diversity, to enrich, embellish and bring accents of human meaning to the urban environment. This meant increasing considerably the size of the designs on which I had been working. I built a lathe that could swing a 7-ft. diameter. I also felt that to fit contemporary architecture, my turnings would require verticals and horizontals. Squares and rectangles are eminently easier to work with if they are to be a part of the building, not just additions hung on the walls at random intervals. The turned face of a square, without being segmented, can make a tile for floor, wall or ceiling. Segmented in halves or quarters and reassembled, face-turned squares suggest many more variations for surface patterns.

Figure 9 shows a square that was turned on both sides, then cut into uniformly thin strips, with the cross section of each strip facing the camera. This solution could become a small grill; turned through 90° it becomes a speaker cabinet or a garden gate. Increase the scale and it might be an 8-ft. screen dividing two spaces; hung horizontally it becomes a false ceiling or the pattern in a terrazzo floor. The potential of such forms can be seen through this example; the precise use will determine the materials, scale and finish.

These speculations relate to faceplate turning and divisions of the square. It is easy to achieve a square from spindle turning as well, by cutting the form through its long axis and rotating the pieces. The piece shown in figure 10 could become a table, or, by increasing the scale, an archway. The form is actually a 6-in. cube set on a reflecting surface, and is from the first generation of this kind of segmenting. The leg is too heavy and the top too thick, but that kind of refinement comes with familiarity.

This essay and the forms that illustrate it document more

Fig. 9: Form exploration, white pine, 13 in. by 29 in., 1974.

Fig. 10: Form, poplar, 6 in. cube, 1974.

than eight years of inquiry. They represent an inescapable curiosity about experience and perception, insight and differentiation. These aspects of creating seem concentric and interwoven, like the annual rings of the tree, which grows to maturity from the center out, from the earth to the sky, taking nourishment from above and below, outside and inside. The ecology of creativity is a delicate balance, usually confusing, often exhausting, but never dull and sometimes exhilarating, for the maker and (one hopes) for the viewer as well. □

Stephen Hogbin is the author of Wood Turning: The Purpose of the Object, *about his experiments during a year's study in Australia. It contains photographs and drawings of the 7-ft. swing lathe he built. (The book, published by Van Nostrand Reinhold in 1980, is now out of print; check libraries or used-book stores.)* The Unknown Craftsman *can be purchased from Books about Wood (RR3, Owen Sound, Ontario, Canada N4K 5N5), a mail-order service operated by Hogbin's wife, Maryann. Hogbin sells his woodturnings through the Aggregation Gallery in Toronto, Canada.*

Fig. 11: Untitled, red oak, 11-in. radius.

The Sketchbook as a Design Tool

by Leo G. Doyle

There is more to woodturning than mere operation of the lathe. Most books and courses deal mainly with technique, devoting little time, if any, to design. The serious woodturner should seek a more equitable balance between designing and actually working on the lathe. I can design many more pieces than I have time to make, so when I do decide to make a thing, I can be very selective. I have accumulated through my own research and teaching a number of approaches to design, which may help other turners develop a more professional design sense. Very few people seem to have a natural ability to design—most of us must learn design techniques just as we learn turning techniques.

The heart of my method is my sketchbook/journal. I treat this book with the same respect as a sharp tool. A way to get started is to sketch the turnings you feel have been most successful. It is best to do this in chronological order, so you can look for consistency or signs of maturity, and evaluate your own progress.

The journal is more than sketches of completed work. It quickly becomes a personal diary of your thoughts and observations. It helps if you can draw well (and to learn I would recommend a course in figure drawing, not mechanical drawing), but good drawing is not a necessity. It is more important just to understand what you commit to paper. A sketch can be redrawn and cleaned up later on; its real value is the development of new ideas. I keep one book at home and one at my shop, because many ideas come while I am turning and I want to document them immediately.

Another helpful procedure is to write lists of turnable projects. These lists should not be just single words, but each idea should be expanded to a variety of applications. For example, think about a turned lamp as a problem in lighting. Soon you would list table lamps, desk lamps, chandeliers, wall lights, reading lights, and so on. Expand the lists and make notes about possible constructions with the woodturning process in mind, then you'll be able to extract a practical list of possibilities for further development.

You will also discover that the popular view of the lathe as a machine with narrow limitations is quite unfounded. You can, for example, beat the size limitation by turning a piece in sections to be joined together. Any limitations that seem dictated by the lathe itself should be taken as a challenge to discover new ways and ideas.

If you are serious about learning to design, there are more exercises to try. We get information about form from our eyes. Train yourself to see the round shapes in nature as well as those made by man. Everywhere you look you can find round shapes that can increase your understanding of form, line and proportion. Pay attention to work in other media with round forms; in particular, to books on ceramics. Pay attention to sculptural possibilities and practice drawing by sketching nonfunctional forms. Many times this approach can loosen up new directions, because the shapes can be used for functional pieces as well. Try to design multi-functional forms and bring your findings back to traditional turnings.

The woodturning process is a perfect model maker. You can quickly try an idea at small scale in whatever wood is handy. This allows the close study of forms for larger projects, as well as plain development of ideas. Look at your models from different angles, on the floor, walls, ceiling, upside-down. Elongate shapes you like, or compress them. Cut the shapes apart and reassemble them, keeping track of your thoughts in the sketchbook. Try adding bits of other materials, or colors. Learn to see the negative space around the ob-

From *Fine Woodworking* magazine (March 1980) 21:60-61

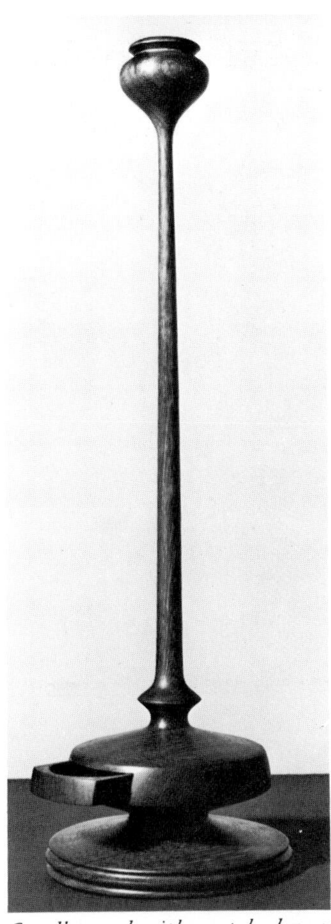

Candlestand with match drawer stacked into blank before turning, padauk, 5 in. by 18 in.

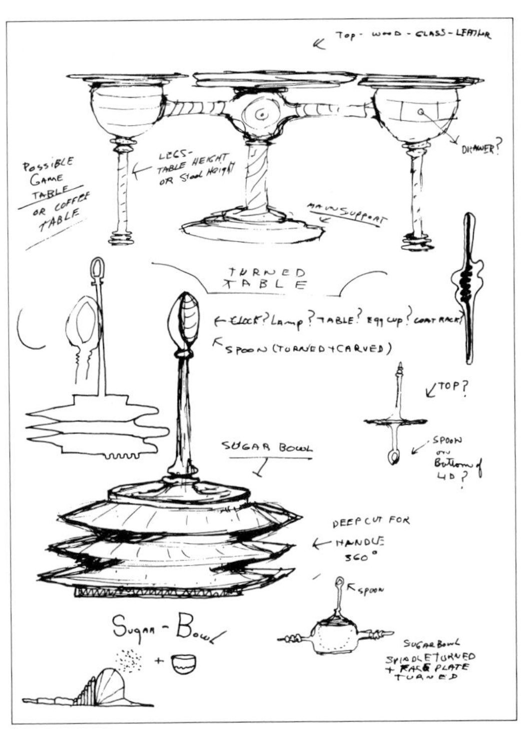

Sugar bowl, purpleheart, 3½ in. by 7 in. high. Spoon is carved into lid. Sketchbook page documents author's idea.

Vanity, purpleheart, 5 in. by 10 in., is designed for a desk. Inverted, the mirror becomes a lid.

Sugar bowl, purpleheart, about 7 in. high.

ject as well as the positive space which is the object itself, to be sure they are working together. Woodturning is a process of removing wood—think also in terms of what could be added to a piece.

Many beginners tend to overdesign. I try to simplify shapes. I try to use the basic concave and convex shapes the turning tools make as part of the function of a piece. More intricate shapes will come in time, with experience. We seldom hear the word "delicate" applied to woodturning. This quality should be examined closely—the work of many beginners appears heavy, overdone. How thin a spindle can one turn in various woods, and still maintain functional strength?

There often doesn't seem to be much gap between hand-turned work today and good mass-produced work. Yet there are many techniques a hand-turner can use that will separate his product from industrial production, especially in the areas of split and off-center turning. Does it really make sense to do twist-turning when that process has been mastered by machines? I feel there are too many other possibilities to explore.

Turning can also be a way of highlighting the grain of carefully selected wood. But turners should be careful when combining different woods, because the color contrast can distract from the form. Successful combinations of different woods can require as much thought as the actual design of the object. Many times projects would be much stronger if the turner had used just one type of wood, or added a second variety only for subtle changes.

Woodturning is usually considered an adjunct to other methods of woodworking. I have made it the basis of all my work, and I try to approach every commission I receive in terms of turning it on the lathe. Other methods of woodworking then come into play to support the woodturning, not the other way around. I like to think my work could easily fit in with all periods of furniture—I do not design to conform to any particular style. I would like my work not only to be used, but also to be appreciated for its own existence. □

Leo Doyle is chairman of the art department at California State College, San Bernardino, and associate professor of woodworking and furniture design.

Photos: Leo G. Doyle

Decorative Turning

Plunging right into a bowl's personality

by Tom Alexander

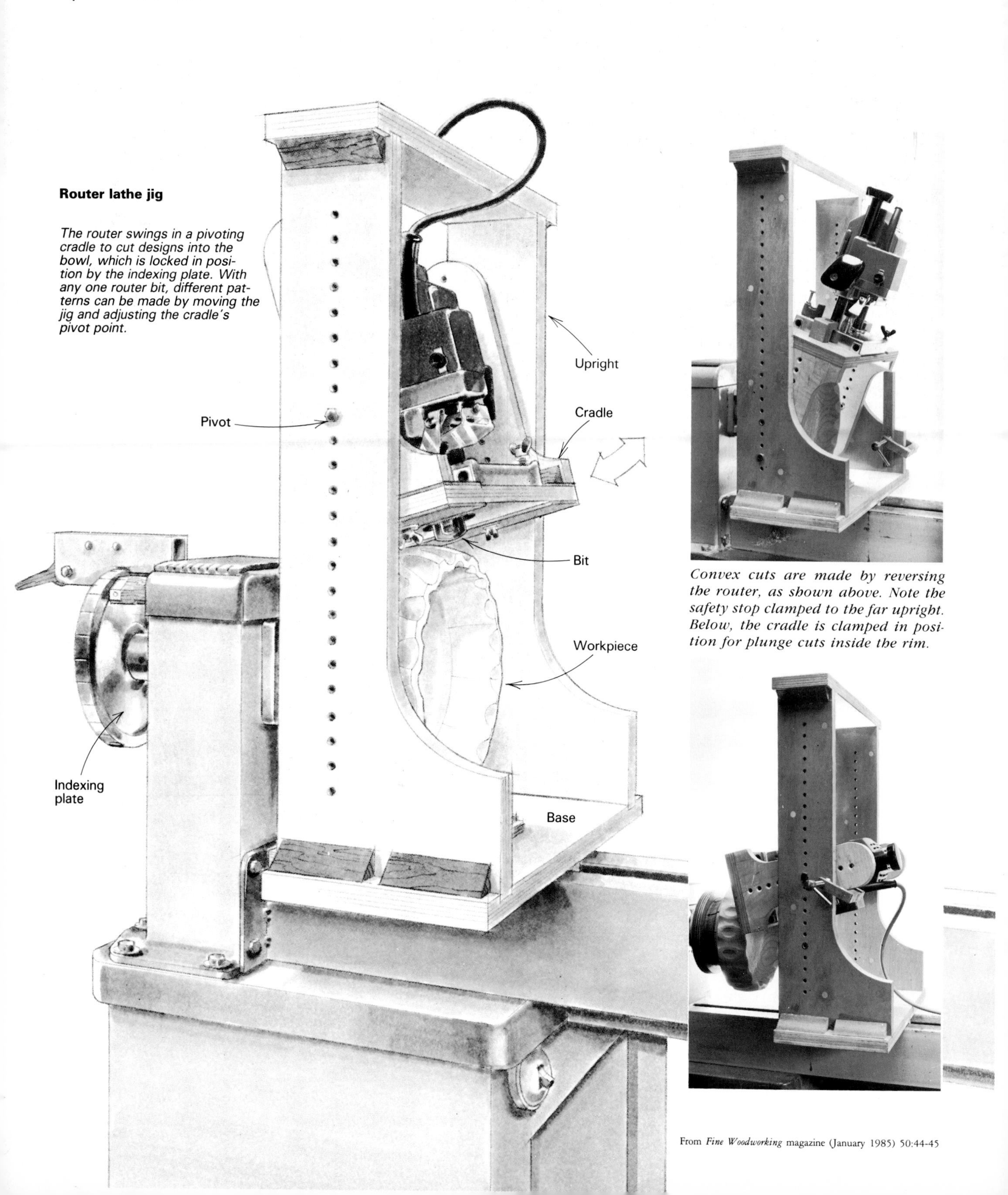

Router lathe jig

The router swings in a pivoting cradle to cut designs into the bowl, which is locked in position by the indexing plate. With any one router bit, different patterns can be made by moving the jig and adjusting the cradle's pivot point.

Convex cuts are made by reversing the router, as shown above. Note the safety stop clamped to the far upright. Below, the cradle is clamped in position for plunge cuts inside the rim.

From *Fine Woodworking* magazine (January 1985) 50:44-45

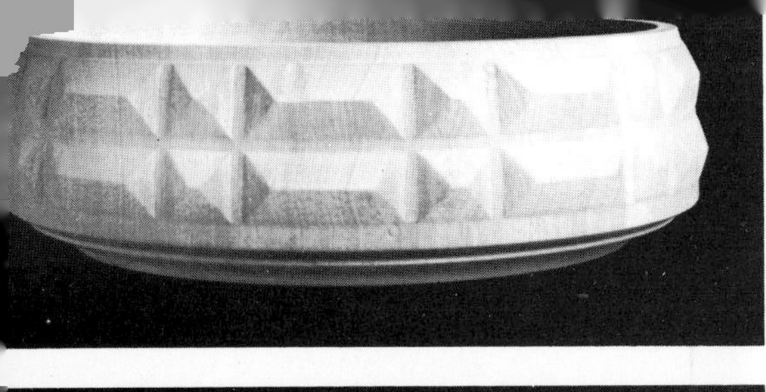

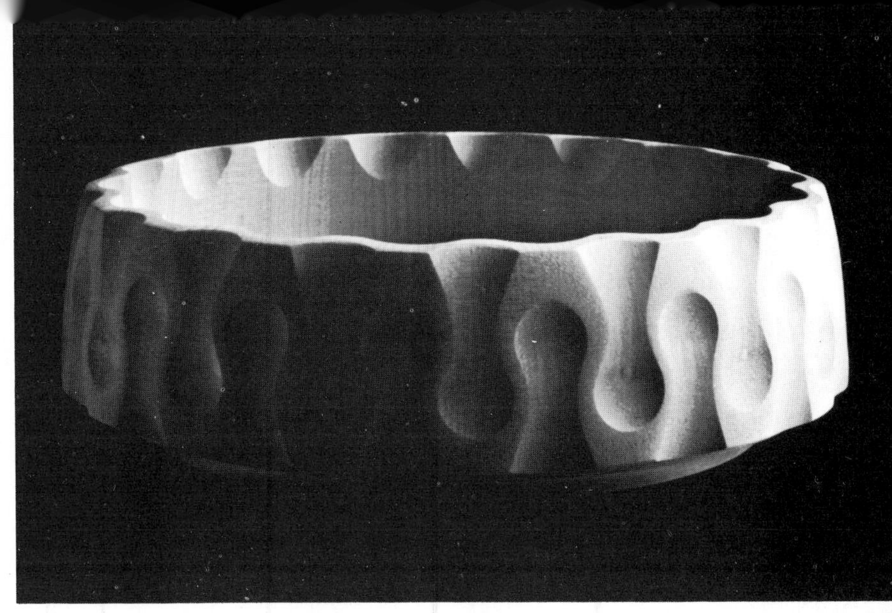

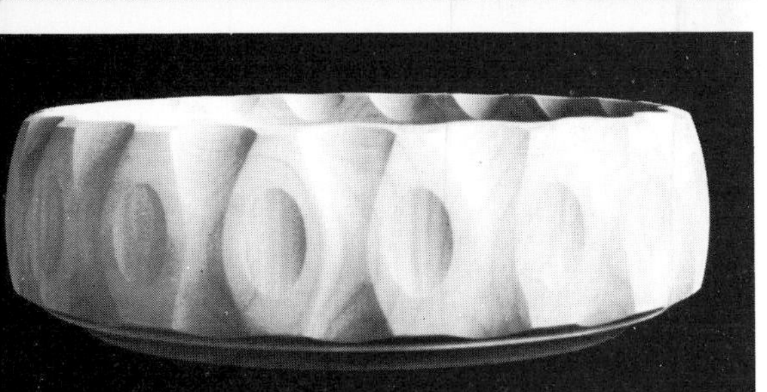

These bowls were all stave-laminated on the same set-up jig before turning, a production process that keeps their basic size and proportions about the same. Yet the incised decoration makes each bowl unique. All cuts were made with the jig on the facing page, using the two router bits shown at right. Alexander grinds his bits from old planer blades and bolts them into a ½-in.-shank mandrel.

The bowls shown here are about 9 in. in diameter, and the patterns on them were sculpted by a router. The process works something like turning on a Holtzapffel lathe: The workpiece is first turned to shape, then the lathe is stopped and the work locked so it can't rotate. A cutting tool, in this case a plunge router in a pivoting cradle, incises an arced groove into the bowl's surface. The piece is then rotated a fixed amount and locked again, and another groove is cut. One or more series of such cuts complete the pattern.

The bowls shown here were turned from stock that was stave-laminated, that is, glued up like a barrel. The technique saves wood, and various jigs make it suitable for production turning. One limitation of using jigs is that all the bowls come out about the same size and shape, but the router's surface treatment gives each one individuality.

You can adapt the methods to any size bowl and to whatever router and lathe you have. I make my own router bits, as shown in the photo above, but standard bits could also be used.

If your lathe doesn't have an indexing plate, you'll have to buy or make one to lock the headstock at various positions. An indexing plate is a perforated wooden or metal disc fastened to the headstock spindle. A pin goes through a hole in the disc and prevents the spindle from turning. The indexing plate can be outboard or inboard—the location depends on the lathe and on whatever locking-pin arrangement is convenient. Some lathes have locking-pin holes drilled right in the pulley. Another option would be to drill holes in the back of the faceplate.

To make an indexing plate, turn a disc from plywood and mark a series of concentric circles on its face. Around each circle, drill a series of holes at fixed intervals for the locking pin (divide the number of degrees, 360, by however many stops you want the circle to contain, then lay out the holes with a protractor so they're evenly spaced). You'll also need a router cradle. As shown in the drawing, the uprights are attached to a base that locks on the lathe ways and can be moved closer to or farther from the work. Inside the uprights, a pivoting cradle holds the router. A row of holes in the uprights allows you to position the cradle at various heights, and a similar row of holes in the cradle itself determines the radius of the arc of the cut. Fine adjustments are made by changing how far the bit extends beyond the baseplate, and most cuts are made in one pass.

In the drawing, the router is mounted inside the cradle and the cradle's pivot point is close to the router base. This arrangement results in a tight concave cut in the workpiece. For the opposite effect, a sweeping convex cut, mount the router as shown in the top photo on the facing page. The bottom photo shows the cradle locked by a clamp—in this setup, the router plunges forward to make patterns inside the rim. If your router doesn't plunge, you can make inside cuts by arranging the pivots so the router's swing is within the bowl instead of outside it, or by making a router cradle that slides rather than swings.

These variables, in combination with choice of bit, allow great versatility. In fact, it's unlikely that you'll ever make two bowls that look exactly the same, however hard you try. I try to visualize each cut before it's made, swinging the cradle to see the arc and sometimes substituting a dowel stub for the bit so I can better see its path. Even so, minor changes in depth of cut or in the profile of the workpiece add up to subtle differences from one bowl to the next. These surprises, fortunately, are usually pleasant ones. One final tip: It's a lot easier to deepen an existing cut than to try to make it a little shallower after the fact. So plan your cuts carefully. If you do go too deep, the only remedy is to turn down the whole bowl a little. □

Tom Alexander turns spinning wheels and bowls in Ashburton, New Zealand. Photos by the author.

Drawing: David Dann

Laminated Turnings

Making bowls from stacked rings, bottles from tall staves

by Garth F. Graves

A woodturner wishing to make large bowls or cylinders will be fortunate to find a suitable piece of premium stock that has been spared from being reduced to veneer, resawn into milled boards, or subjected to the sculptor's chisel. The alternatives are unseasoned blocks that are rough-turned, set aside to dry, and final-turned again (see "Green Bowls" by Alan Stirt on pp. 21-23), or shells laminated from standard stock. Turning forms may be built up from an unlimited selection of choice, seasoned hardwoods. Premium stock may be laminated into a form and size that accommodates the most ambitious project, and design opportunities in size, patterns, shapes and applications are limitless. Laminated shells are as strong as pieces turned from solid blocks, and lathe time and material costs are significantly lower.

Turning blanks may be built up horizontally, by stacking rings, or vertically, with beveled staves. Shallow containers such as bowls and trays lend themselves to stacking, while tall, slender forms usually dictate the assembly of vertical staves. Added design interest can be achieved by contradicting the common logic—there is no reason that stacked rings cannot form a tall cylinder, or that shallow, wide bowls cannot be made from vertical staves. Constructing the blanks gives the craftsman great control over the final form; subtleties often transform a nice turning into something special.

The technique of stacked rings is based on overlapping concentric rings, cut at an angle from a single piece of 1-in. surfaced stock. Experimentation with the concept can produce a fairly wide range of shapes. The surface pattern of the finished piece can be controlled to some degree, and the pattern can be accentuated by inserting alternating pieces of contrasting woods at the rim, the base, or in the wall of the bowl.

The geometry of cutting all the rings from a single source board somewhat limits the shapes obtainable (diagram on opposite page). The concentric rings, cut at a 45° angle, are laminated to form a hollow conical blank for turning. The angle of cut, the width of the rings and the thickness of the wood govern the cross section of the blank. For this basic method, any increase in depth will proportionately increase the diameter. Variations of these parameters will change the assembled profile. The practical limits are quickly reached and for more scope two source boards become necessary.

Using two or more boards for alternating rings increases the possibilities. Inserting straight-walled rings between the angle-cut rings makes the form taller and reduces the slope of the wall. But I've found that the best method is to alternate diagonally-cut rings from two source boards. This reduces the slope of the wall; the thicker walls allow greater design variation. Further variations include leaving a wide ring at the top for a flare, a flange or a handle. The pieces from which the rings are cut may be segmented and laminated in many ways or built up to almost any size.

For proper bonding, the boards used for stacking should be milled and planed truly flat and evenly thick, free of ripples, valleys, pecks and checks. The circles can be cut by hand or with a jigsaw or sabre saw. But before cutting, draw a random series of concentric circles on the underside of the stock to aid alignment during assembly. The saw blade must have a starting hole, but instead of a single large hole, which would remove too much wood, drill a series of small holes in line with the arc of cut and at the angle of cut.

Keep the grain parallel from ring to ring, so seasonal movement won't break the bowl. Allow glue joints to set under pressure, attach to a faceplate, and you're off and turning.

I've seen many otherwise fine turnings marred by screw holes in the bottom, the result of attaching the faceplate directly to the wood. Glue a square of scrap to the turning blank, separated by a single sheet of newspaper. The paper will separate when a knife blade is forced between the scrap and the finished piece, and the faceplate may be securely screwed to the scrap piece.

Laminated rings are only one way of building up blanks for turning. The assembly of a number of wedge-shaped staves to form a cylinder opens more opportunities for project design. A wide variety of shapes and sizes are possible—from large cylinders to bowls, buckets, or salad sets. All can be produced from standard milled stock of premium woods, but making cylinders of vertical staves requires care. Clean, true cuts are essential and provision must be made for fitting ends to the cylinders. The same principles apply to compound-angle forms that would result in conical blanks.

Success comes from properly joining the staves; the angles are critical. The table at right contains information on various final diameters, the number of staves required, the angles of cut, and the outside width of each. It also gives the general mathematics for any size cylinder. The thickness of the stock and the number of staves will govern the wall thickness of the finished piece. The more staves, the more circular the blank will be, and therefore the greater the usable thickness.

After the staves are cut to the proper angles they are assembled dry into cylinder form and checked for fit. Some adjustment is possible by trimming angles or adding segments, to compensate for miscalculations in the cutting angle. Remember, any error is multiplied by the two surfaces and by the number of staves used.

The photo sequence on page 78 highlights the application of this technique. Note that thick stock was required by the widely varying diameters of the finished pitcher; 1-in. stock would be prepared in the same way. When the shape restricts access to the inside, as in the piece shown, I find it helpful to diagram the cross section. I use inside calipers or dividers to measure along the center line at 1-in. increments, then I dia-

(please turn to page 78)

Garth Graves, of San Diego, is a designer and prototype woodworker. He used to be an aerospace technical writer.

From *Fine Woodworking* magazine (Spring 1978) 10:70-72

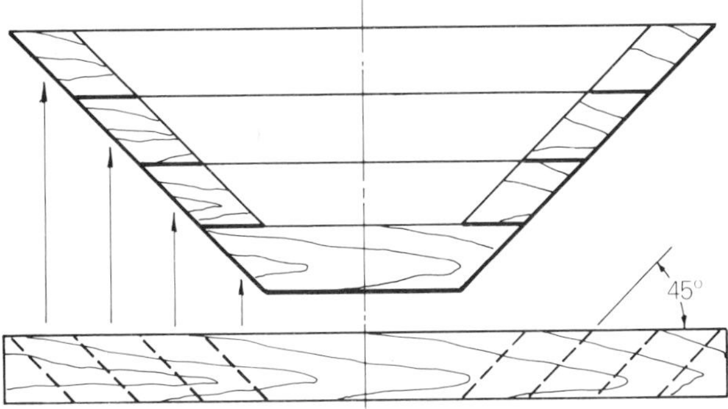

Cross section above shows how a single source board, cut at 45° into rings whose width equals the thickness of the stock, can be stacked into a conical turning blank.

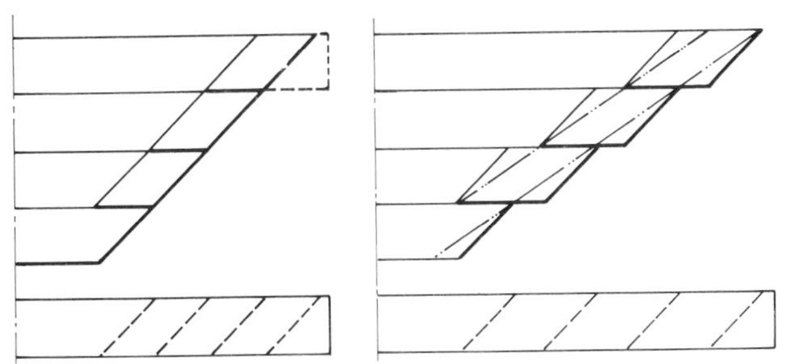

The top ring may be cut wider, left, to turn a bowl with a flaring rim. Increasing the ring width to 1½ times the stock thickness, right, permits a slightly wider bowl but possibilities are limited.

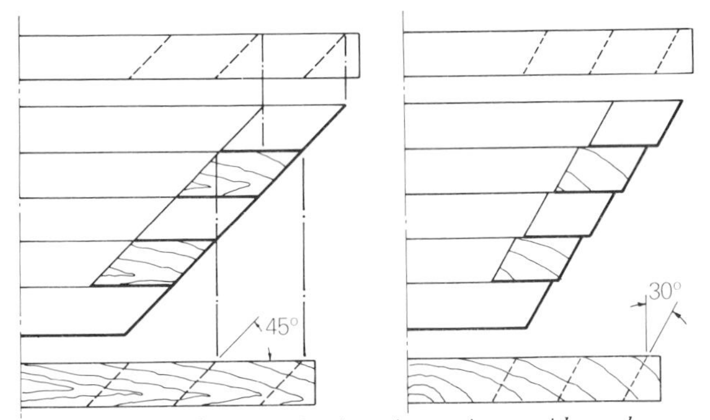

Two source boards, cut at 45° into rings twice as wide as they are thick, allow much more variation in shape. Note that the alternating rings are offset by half of their width. To make a steep bowl with thick walls, cut two boards at 30° into rings 1½ times as wide as they are thick, as at right.

Bowls are turned from stacked rings of ¾-in. teak.

Author turns woodenware in teak from staved cylinders; decanters are about 12 in. high. Chart below relates cylinder diameter, number of staves and stave width. Half-angles are the amount of saw-blade tilt from the vertical when cutting staves.

θ Angle	N Qty. Reqd.	Half-Angle Cut	STAVE WIDTH W_1 for given diameters 3 in.	5 in.	7 in.	9 in.	10 in.
90°	4	45°	3	5	7	9	10
60°	6	30°	1¾	2⅞	4	5¼	5¾
45°	8	22½°	1¼	2⅛	2⅞	3¾	4¼
30°	12	15°	¾	1⅜	1⅞	2½	2¾
15°	24	7½°	⅜	⅝	1	1¼	1⅜

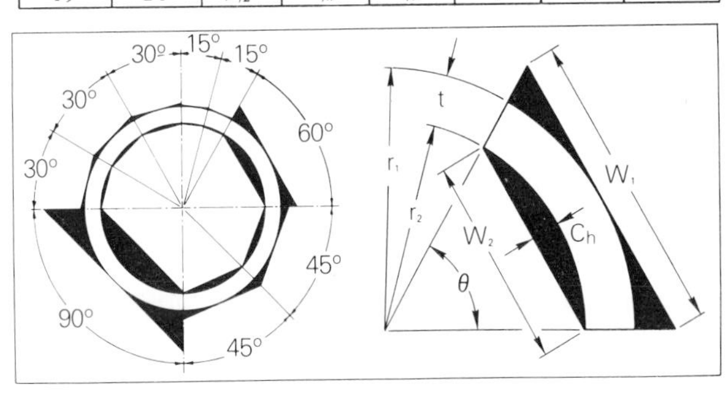

The diagram above left shows a few of the regular divisions of a circle and the relationship between the number of staves in a cylinder and the stock thickness required to produce a given wall thickness. The enlarged section through a stave, right, defines the terms for the following equations. Once you decide what you want to make, the math gives minimum dimensions. Keep angles precise but keep outside width and thickness fat, to have wood for working.

Here, N is the number of staves and θ is the included angle of each; thus, $\theta = 360° \div N$; r_1 is the outside radius of the finished cylinder and r_2 its inside radius; t is the wall thickness and thus $t = r_1 - r_2$; W_1 is the outside width of a stave, W_2 is the inside width; C_h is the chord height and thus stock thickness $T = t + C_h$.

If you have decided on the number of staves and the radius of the cylinder, solve for the width (W_1) of each stave:

$$W_1 = 2r_1 \tan \frac{\theta}{2}$$

For example, if you want a cylinder that has an outside diameter of 8 in., and it will be made of twelve staves ($\theta = 30°$), then, $W_1 = 2 \times 4 (\tan 15°) = 8 \times 0.2679 = 2.14$ in. Set the table saw at 75° (90° minus the half-angle, or 180° ÷ N), and cut each stave so it is at least 2.14 in. on its outside face.

To find the minimum thickness of the stock (T) for a given wall thickness (t), first solve for the chord height:

$$C_h = r_2(1 - \cos \frac{\theta}{2})$$

In the previous example, if the wall is to be ½ in. thick, then $C_h = 3.5\ (1 - \cos 15°) = 3.5\ (1 - 0.9659) = 0.119$ in. Since $T = t + C_h$, $T = ½ + 0.119 = 0.619$ in.

To find the width of the inside face of a stave, solve:

$$W_2 = 2r_2 \sin \frac{\theta}{2}$$

In the previous example, $W_2 = 2 \times 3.5 \times 0.2588 = 1.812$ in.

gram the measured diameters onto a cross-sectional view. This way I can determine the wall thickness anywhere. A design change in the final stages need not conform to the inside profile, as long as the wall doesn't get too thin.

A cylinder formed of staves doesn't expand and contract the same way as one turned from a solid block. Shrinkage will occur evenly around the circumference and across the wall thickness but the form won't become ovoid. If shrinkage is expected, any lids should be fitted loosely.

Segmenting permits variations that would not be considered if the cylinder were turned from a solid piece. I don't hesitate to embellish a segmented piece with handles, pouring spouts or whatever the design requires. One container shown earlier includes a spout. Prior to assembly, I scored the stave where the spout would go about one-half the way through, and did not glue the portion to be replaced. After the final turning, this piece was easily cut away. A rough-shaped spout, cut to the same half-angle, was glued in place for final shaping. The handle, although attached separately, could have been added in the same way. □

1. *Shopsmith saw table is tilted to half-angle; hollow-ground blade produces good gluing surface.*

2. *Check fit dry, then clean surfaces and glue. Author supplements cord wrap with vise-grip chain clamps or band clamp. Keep the assembly vertical.*

3. *With the top turned true to the sides, the cylinder is screwed to a faceplate. Screw holes will be turned away later.*

4. *First the outside diameter is roughed, then the bottom and lower inside diameter are finished to receive the base.*

5. *Cylinder is removed from lathe and replaced by stock for base. Author usually turns a rabbet so base will plug in snugly.*

6. *Tailstock feed applies pressure for gluing cylinder to its base.*

7. *Now the inside top and the whole outside of the cylinder may be turned to their final shape.*

8. *Use the lathe bed as a holding fixture for adding handles, spouts—whatever the design requires.*

Compound-Angled Staves

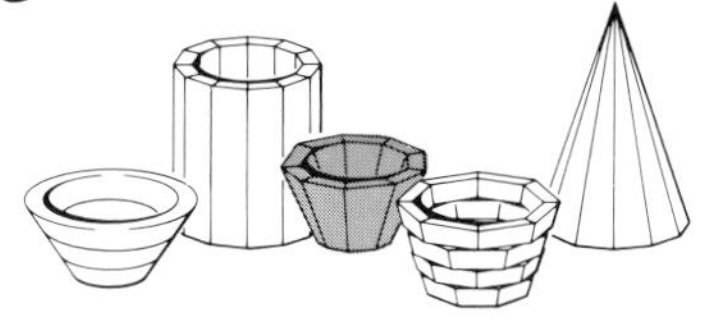

The previous discussion covers building up conical turning blanks from stacked rings, and cylindrical blanks from beveled staves. The next logical step is to make a bowl-shaped blank from staves that are both tapered and beveled.

Robert M. Hewitt, 47, a structural engineer from Mechanicsburg, Pa., has developed a simple method for cutting the staves with his radial arm saw, and for clamping them together with a nylon cord.*

Hewitt's method assumes you have already determined the number of staves (N) in the bowl, and the angle α between the side of the bowl and the table. To cut the segments on a radial arm saw, you'll need to know angle a, the bevel setting of the saw blade and arbor, and angle b, the miter setting of the saw arm. The formulas are:

$$\text{angle } a = \frac{180°}{N} \sin \alpha$$

$$\text{angle } b = \frac{180°}{N} \cos \alpha$$

For example, a bowl with 12 staves sloping at 60°:

$$a = \frac{180°}{12} \sin 60° = 15° \times 0.866 = 13°$$

$$b = \frac{180°}{12} \cos 60° = 15° \times 0.5 = 7.5°$$

If he wants the grain to be vertical in the finished bowl, Hewitt selects a wide board at least an inch thick and crosscuts it into strips whose length equals the height (h) of the finished bowl plus an allowance for cutoffs and for truing up the bottom before gluing it to the base. He bevels the edges to the same slope α as the staves will make with the base, and saws two shallow kerfs near the edges, as shown in the drawing. The kerfs will hold the nylon cord during glue-up.

Next, with the saw blade tilted to angle a and the arm at angle b, he makes the first cut at one end of a strip of stock. Use an adjustable drafting triangle to set the saw accurately. Next he locates and fixes a stop to the table so that the dimension W is the width of a stave at the base of the bowl. This can be guessed, or calculated with the equations given on page 81. He flips the stock over, indexes it against the stop and makes the second cut, producing one of the 12 staves required. Flipping stock again, he makes the next cut, and so on until he has all 12. Hewitt leaves the saw set up, so he can adjust one stave to compensate for error and close the bowl.

Assembling tapered staves is tricky, and Hewitt has a tricky solution. He writes "Lay the tapered staves together on a sheet of paper, face down. They will form a segment of a circle, like a pie with a large piece removed. Outline the assembled pieces, and number them so they can be replaced in the same order. Apply contact cement to the paper and the outside face of the staves between the cord kerfs, trim the paper to the outline and cement the staves to it. This will allow you to pick up all the staves at once, close the bowl and see how they fit." One stave may have to be adjusted to make all the joints close tightly and it is usually easiest to cut a new one at a slightly different bevel angle.

Hewitt continues, "When the staves fit, lay the bowl out flat again on the bench, cut two lengths of nylon cord and knot them into loops a bit larger than the circumference of the closed bowl. Apply glue to all the mating surfaces, close the staves to form the sides of the bowl, slip the nylon cord into the kerfs and use a dowel to twist the cord for clamping pressure." When the glue is set, Hewitt glues the top of the bowl to a disc of plywood and centers it on the lathe. Then he turns the bottom of the blank true, and glues on what will be the bottom of the bowl. He now can turn and finish the bottom and outside, mark the center, reverse and remount, and part off the plywood. The bowl is completed in the conventional manner. Sometimes he puts a contrasting veneer between staves, for visual interest. He recommends finishing with four coats of satin urethane varnish, rubbed with pumice and oil. This seals the wood completely and prevents expansion and contraction of the wood segments. —*J.K.*

*Hewitt's method was presented at the March 1977 Woodturning Symposium in Newtown, Pa. Organizers Albert LeCoff and his twin brother, Alan, and Palmer Sharpless ran a series of these symposia, finally calling it quits after the tenth one, conceived as a climactic recapitulation of the series.

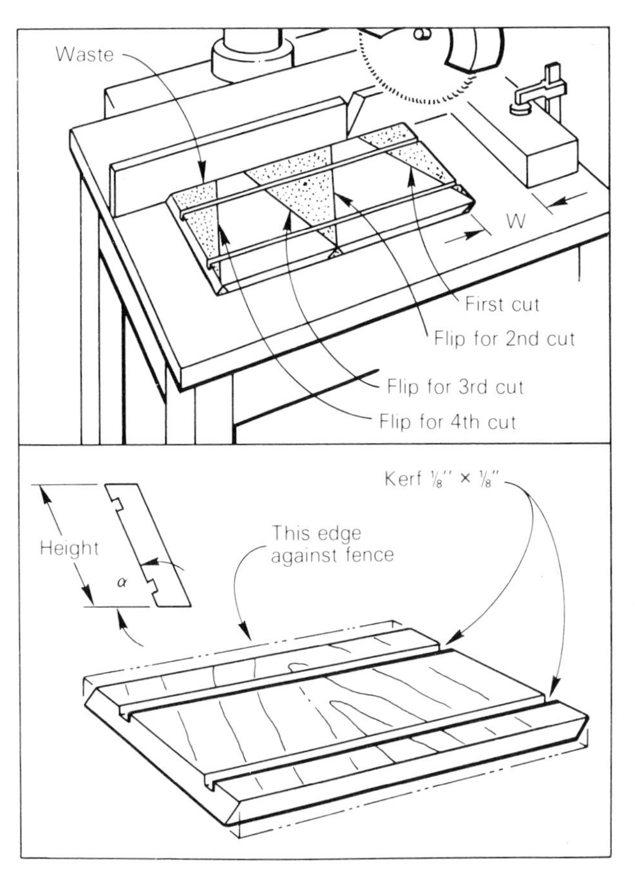

Blank is glued to waste disc so foot may be trued and bottom attached. Right, cherry bowl with horizontal grain and walnut base.

From *Fine Woodworking* magazine (Spring 1978) 10:73

Rings from Wedges

by Asaph G. Waterman

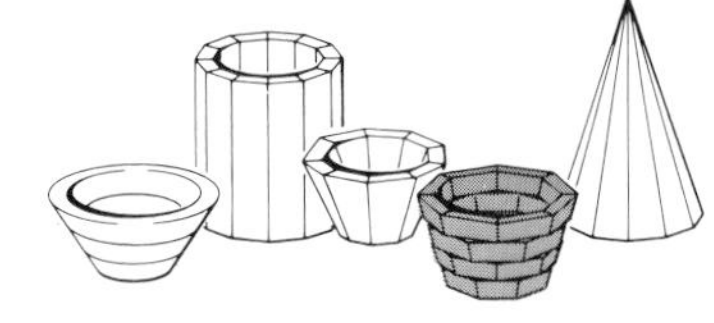

(Now imagine that a cylinder is made from very short vertical staves—no taller than the thickness of a board. The staves become wedges, and the cylinder is squashed into a ring. Several such rings can be stacked up to form the blank for a bowl. Asaph G. Waterman of Camillus, N.Y., has devised a tablesaw jig for cutting wedges, and a plywood-and-angle-iron jig for gluing them together.)

There are several advantages in using this technique: The wood need not all be the same thickness, so scraps left over from other projects can be used up; wood with nail holes and other imperfections can be used with minimum loss; no end grain has to be turned, especially important in soft woods like butternut or sumac; and striking effects can be obtained by gluing contrasting pieces of veneer between the wedges.

Accuracy in assembly is very important. The jigs I will describe are for making eight-sided (octagonal) rings, but the same principles apply to jigs for any number of wedges.

The two sliders for the table-saw jig may be made of steel, aluminum or hard wood. They should fit your table-saw grooves accurately, but must slide smoothly. Make the body of the jig, as shown in the drawing, of any stable wood 5/4 thick. The angle, 67½° for an octagonal ring, must be accurate because any error will be multiplied by 16 in the finished blank. The triangular wedges for the first layer come to a point, to fill the bottom of the bowl, but succeeding rings must be wider to allow the bowl to flare, and cut off (truncated) at the point to save turning work and avoid waste. The pointer on the jig is used to gauge the width of the truncation. Face the working surface of the jig with coarse carborundum cloth to keep the wood from sliding.

The assembly jig consists of an octagonal plywood base to which are screwed eight 2½-in. lengths of angle iron, drilled and tapped for tightening bolts. I have jigs in two sizes—one an octagon 11 in. from face to face, with sides about 4½ in. wide; the other 14½ in. from face to face with 5¾-in. sides. I use 1-in. by 1-in. by ⅛-in. angle iron. Don't use aluminum angle because the threaded holes won't stand continued use. One side of each piece of angle iron is drilled in its center and tapped for a 5/16-in., 16-pitch machine screw or cap screw. The other side is drilled ⅜ in. from each end and countersunk for ¾-in. flathead wood screws.

Because the layers have varying diameters, I use spacer blocks in sets of eight between the tightening screws and the wedges themselves. My jigs will make a bowl 11 in. in diameter; a larger bowl of course requires a larger jig.

To make a bowl, I first saw eight pieces that come to a point, turning the board for each cut. Make sure the pieces fit, lightly sand each edge, and, since the main problem in this work is getting the points to meet exactly, avoid it by sanding the points off flat. Later I drill out the center of the octagon with a tapered bit and turn a tapered plug to fill the hole. I usually run grain of the plug parallel to the grain of the layer itself. Use the small assembly jig with appropriate spacer blocks to glue all the wedges together at once. Protect the jig surface with waxed paper, wipe off the excess glue and let set overnight. If the pressure of the screws forces a wedge to rise, use a C-clamp to force it back down. When the glue has set, fasten a faceplate to the layer and take a thin cut on the lathe to smooth and true the surface.

I make the wedges for the second layer the same way, except I move the pointer on the sawing jig about an inch out from the blade. When the glue has set, fasten this layer to a faceplate, placing the screws near the center so the holes will disappear during turning. Smooth one side and use C-clamps to glue both layers together, staggering the glue lines. Now true both surfaces on the lathe.

At this point a two-layer bowl can be turned or a third layer may be added. To make the larger third layer, move the arrow about 1½ in. out from the blade. You can add as many layers as you like, to get as deep a bowl as you want. □

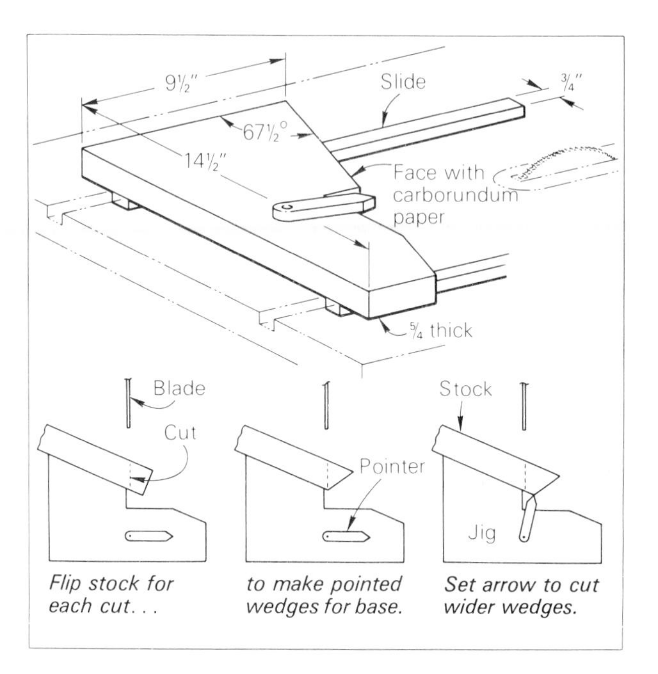

Assembly jig consists of angle iron bolted to plywood plate, drilled and tapped for cap screws. Use spacer blocks between screw and work.

From *Fine Woodworking* magazine (Spring 1978) 10:74

Staved Cones

The general mathematics

by Thomas Webb

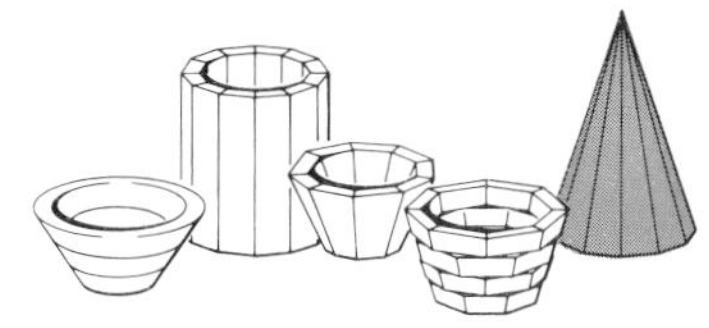

An equal miter joint for a rectangular, box-like construction will result from cutting adjoining edges of the stock at 45°. But what if we need to miter a shape that isn't rectangular? What if, for instance, the shape has seven similar sides that must lean in (or out) rather than standing parallel to one another?

Such non-rectangular forms can be thought of as sections of faceted cone-like shapes. By thinking of forms you wish to make as sections of cones, you can determine the geometry of the flat pieces needed to make those forms. You need only specify the height and base dimensions of the imagined cone, along with the number of sides you want it to have. The formulas will then tell you what shapes to cut to produce the faceted cone shape. Further alterations of the size of these pieces can produce any section of the specified cone. Combining sections of different cones can produce an infinite variety of three-dimensional shapes.

Some of the formulas look complicated, but with a table of trigonometric functions it is fairly simple to do the computations; they can be done in minutes on a calculator with trig functions.

A right cone has a circular base and is symmetrical around an axis running through the center of the base to the tip. The axis is perpendicular to the plane of the base; the length of the axis from base to tip is the cone's height (h_c). A right cone can be described in terms of its height and the radius of its base (r_b); α designates the angle between the surface of the cone and its base.

A right cone can be constructed from a flat sheet of flexible material such as paper or thin metal. The shape to be cut from the flat stock is a circular disc with a wedge removed, and the straight edges of the wedge are radii of the disc (r_d). When the cut radii are pulled together, a right cone is formed; the center of the disc becomes the tip of the cone, the perimeter of the disc becomes the base and r_d becomes the length of the side of the cone.

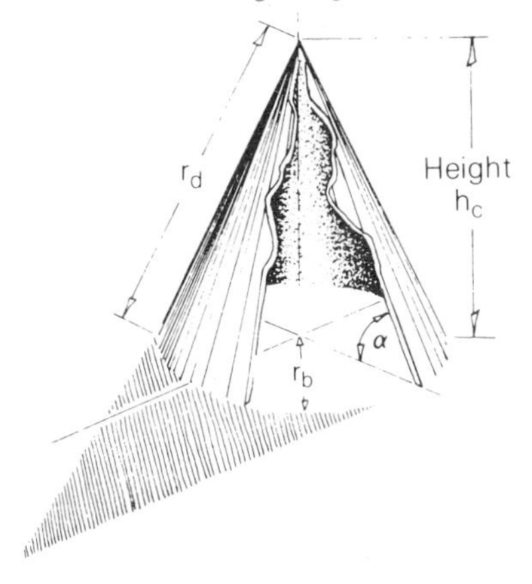

If you know the height and radius of the base of a right cone you want to construct, the following formulas specify the size of the disc you'll need and of the wedge to be cut from it.

$$r_d = \sqrt{r_b^2 + h_c^2}$$

$$\theta = 360° \frac{r_b}{r_d} = 360° \cos \alpha$$

Since θ equals the number of degrees remaining in the disc after the wedge is removed, $(360° - \theta)$ equals the angle of the wedge itself.

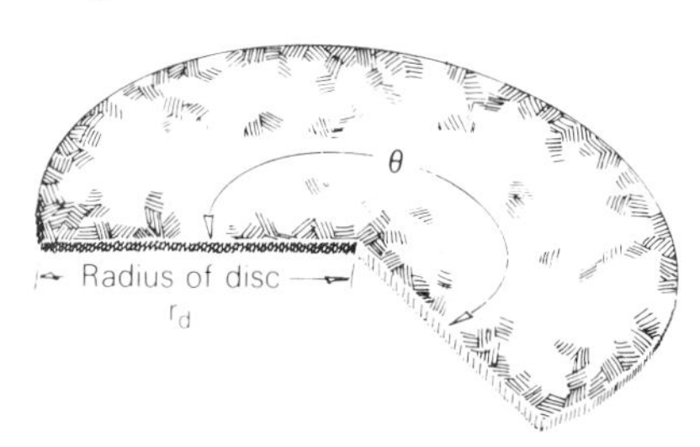

In some circumstances it is useful to specify the base angle α rather than the height of the cone. In this case first solve for h_c using the following formula:

$$h_c = r_b \tan \alpha$$

then apply the previous formulas.

A right cone can be approximated with thicker, less than flexible materials such as wood by cutting compound-angled staves and assembling them around a central axis. The result is a faceted "cone;" the more staves used, the closer the approximation to a true cone.

It helps to imagine that the faceted cone just fits inside an actual cone of similar dimensions. In this way we can see the important dimensions h_c and r_b as they relate to the staves to be made. Follow these steps to determine the size and shape of the staves.

First, decide on the height (h_c) of your cone, the radius of its base (r_b) and the number of staves (N) you want to use to make it.

Compute r_d, which is equivalent to the length of the edge of a stave, by:

$$r_d = \sqrt{r_b^2 + h_c^2}$$

Compute the width at the base (w) and the height (h_s) of each stave by:

$$w = 2r_b \sin \frac{180°}{N} \qquad h_s = \sqrt{r_d^2 - \left(\frac{w}{2}\right)^2}$$

Knowing the height of the stave (h_s) along with its width (w) at the base, you can lay out on your stock what will be the exterior surface of each stave. Remember that the height of a stave is measured along a line that bisects its base at a right angle.

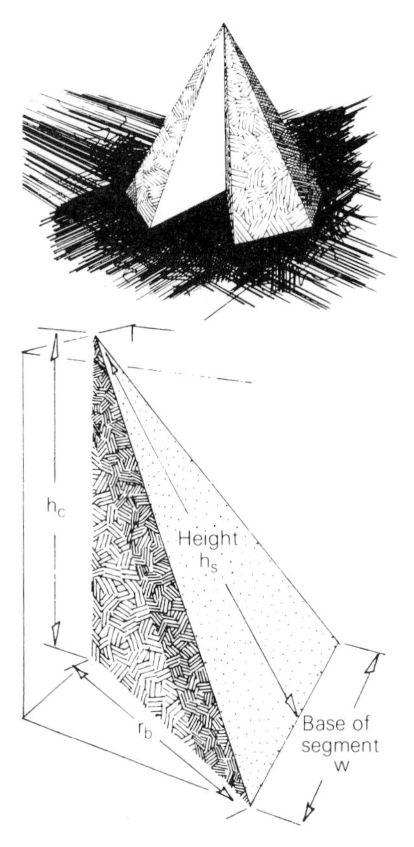

To determine the saw setting for cutting the miter angles on the sides of the segments, first calculate angle α for the shape you are making:

$$\tan \alpha = \frac{h_c}{r_b}$$

Then use this formula to find angle Ω, the table-saw setting for the side cuts:

$$\tan \Omega = \sin \alpha \tan \frac{180°}{N}$$

If you are making a complete faceted cone shape, you may want to have a flat bottom on it. First calculate the base angle β relative to the exterior surface of the segment, from the formula:

$$\tan \beta = \frac{h_c}{r_b \cos \frac{180°}{N}} = \frac{\tan \alpha}{\cos \frac{180°}{N}}$$

Then find the saw setting for the base cut by subtracting angle β from 90°.

Remember that the sides and bottom of a stave converge; the interior surface consequently is a scaled-down version of the exterior surface. □

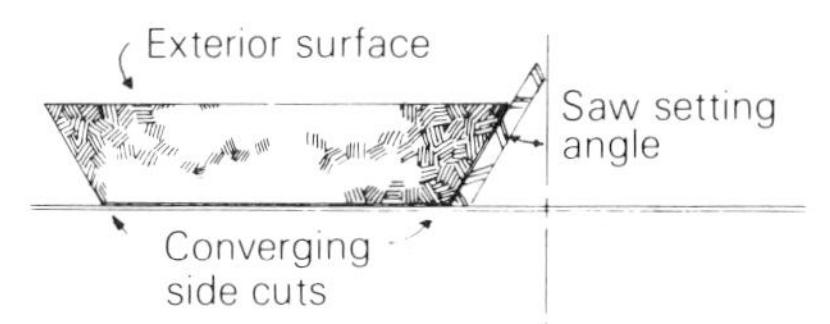

Tom Webb, of Akron, Ohio, is a sculptor and assistant professor of art at the University of Akron.

From *Fine Woodworking* magazine (Spring 1978) 10:75

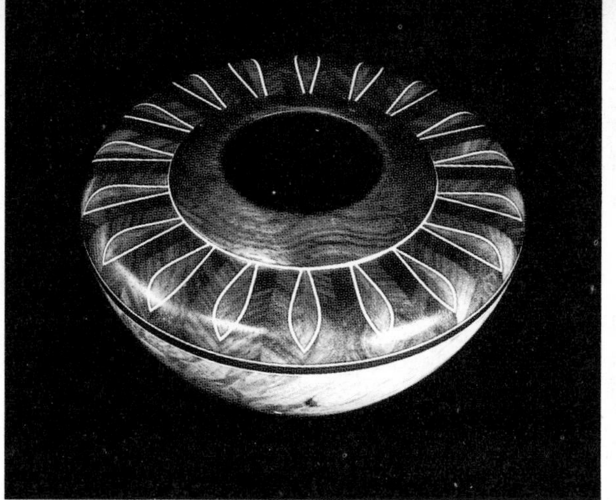

Subtle or showy, the colors and textures of wood, burl and veneer interplay in authors' geometric designs. Glued up from bands of wedge-shaped segments, the lathe-turned bowls shown play on Indian-pottery forms and patterns from the American Southwest. Turnings shown range in height from 4 in. to 9 in. with an average wall thickness of ⅛ in.

Segmented Turning

Redefining an old technique

by Addie Draper and Bud Latven

While exploring the art of woodturning we rediscovered and refined an old turning technique known as segmentation. This process involves gluing various shapes and colors of wood together then turning them to create lively, infinitely-variable designs. The method is time consuming and often complicated but worth it for the richness and diversity of designs it makes possible.

We've always savored the beauty of burl and we find that the regulated crispness of segmented designs contrasts nicely with the unpredictable burl figure. Once a year we go on burl-hunting trips to the West Coast. Some of our favorite species include walnut-root burl from the Sacramento valley, lilac burl from southern California and maple-root burl from western Oregon. The challenge is to devise patterns that complement each burl's unique character.

It's hard to say why we combine certain woods or why a particular pattern complements a particular shape. What works aesthetically and what doesn't is largely subjective but color, figure and density are the main things to consider in planning this kind of turning. If a burl's figure is subdued, a segment band of brightly contrasting wood—rich, red paduak on a field of ash—will liven up the piece. For accent, we might work contrasting or complementary-colored veneers into the pattern. As a rule, we avoid highly-figured woods in the segment band because too much figure is distracting. Woods of similar density turn and sand more easily. If you juxtapose a soft wood next to a hard, dense one, the finished surface may have an uneven feel. This isn't necessarily a liability—the contrast can produce some interesting tactile qualities—but it's something to consider when choosing woods.

The shapes we turn have evolved from many sources. Living in the Southwest, we've been influenced by the indigenous Indian pottery. Though these shapes were developed in clay, they have a directness and simplicity of line that translates nicely to woodturning. Other cultures have provided rich inspiration as well. The jar-like bowl in the top left photo, for example, owes much to the Greek hydria.

Whatever the shape, we usually begin a bowl by drawing it full-size on graph paper. We start with an elevation view including the segment pattern, then make a full-size plan view of each segment band, as shown in the drawing on p. 84. This is a critical step because it allows us to refine the shape and plan the segment pattern in minute detail. We can then measure the various angles and sizes of each segment right off the drawing instead of puzzling it out mathematically. We've found that it's a lot simpler to measure in millimeters and centimeters when working at this scale.

The basic building block for our designs is a segment with a truncated-wedge shape. A few of our bowls are turned entirely of segmented sections, but most consist of a burl blank onto which we glue one or more segment bands that make up the desired pattern. The segment patterns can become quite complex. It's impossible to describe in this article how we make every one, but by experimenting with the basics explained here, you'll be

From *Fine Woodworking* magazine (September 1985) 54:64-67

able to figure out the more complex patterns and invent new ones. We've developed different types of segments and several techniques for combining the segments into patterns. Solid-block segments, and slant-line segments, both shown on p. 84 are the two basic segment types. Solid-block segments are cut from solid wood. Slant-line segments are sliced from a glued-up sandwich of multi-colored woods and veneers so that the laminates form a diagonal stripe across the segment face, as shown in the drawing. The apparently curved lines of holly veneer in the bowl, facing page, center, are simply a variety of straight slant-line segments. The lines appear curved because they traverse the bowl's radiused edge.

Solid-block and slant-line segments can be combined in many different ways, but there are two basic ways that we glue up the segments—single-angle and multi-angle patterns. A single-angle pattern, like a pie cut into equal slices, is made up of segments with the same angle, say 18 segments at 20°. A multi-angle pattern alternates segments with different angles, say 12 segments at 20°, and 12 segments at 10° spaced wide, narrow, wide, etc. Photo #11 (p. 85) shows a simple multi-angle pattern. Whether single-angle or multi-angle, the segment angles always add up to 360°.

By stacking bands of segments we create a multiple-row pattern. The complex Aztec design shown in the photo (facing page, right) is an example of a solid-block, multi-angle, multiple-row pattern.

Once you have drawn the design full-size, you are ready to make the segments. The diameter of the segment band should be 1cm larger than the finished diameter to allow for turning. You can measure the angle and width of each segment at it's heel and toe right on the drawing, or, if you are mathematically inclined, you can calculate the circumference with the formula $C = \pi d$ and divide this by the number of segments if all the segments are the same angle. This will give you the approximate segment width at the circumference. It won't be entirely accurate because it will be the arc of the end of the segment, so the segment itself will be slightly smaller. But since the segments are cut slightly oversize anyway, the measurement you get with this formula will be close enough.

We usually plan to cut the segments so that we don't have to turn end grain. The grain direction should run around the circumference of the turning. We cut the segments 2mm oversize on either side, then sand them to size on the disc sander with an 80-grit disc. Solid-block segments are crosscut from a board on the tablesaw. The sawing process for slant-line segments is two-step, as shown at the bottom of p. 84. Remember that slant-line segments are sliced from a glued-up sandwich of contrasting woods. First, on the tablesaw, we crosscut parallelogram sections from the sandwich. Then we square off the corners of the parallelograms on the bandsaw. For sanding the segments to size, we've tacked a wooden fence to our disc-sander table at 90° to the disc. A wooden push stick (one for each angle) with the end cut to the segment angle feeds the segment into the disc. To ensure consistency, use one segment, sanded to the proper size, as a pattern for the others.

Before gluing the segments to the burl blank, we level the turning blank with a custom-made sanding disc, as shown in photo #4 on p. 85. This 12-in.-diameter disc was machined from ½-in.-thick aluminum by a local machinist. It has a steel spindle with a morse taper that matches the drill press quill. After the segments have been trued up on the disc sander, they're ready to be glued to the burl. With a circular protractor, duplicate the radial "grid" from the full-size drawing directly on the burl blank. Accuracy is very important. Lay the segments on the grid and bring them in tight to check for a proper fit. If there is a total of more than 1mm of slop all the way around, the segments need to be touched up on the sander.

When the fit is right, apply glue to the bottom of a segment and rub it onto the blank moving it back and forth and applying downward pressure. We use Titebond glue for all our laminations. Line up one segment side along a radius line and lay a straightedge along the side of the segment. The straightedge should still follow a radius line on the opposite side of center. Allow the glue to set for five minutes. Glue the next segment and rub it into position against the first, but don't bump the already-glued segment out of position. Continue this process, gluing each segment in place around the circumference of the blank, taking care to align each segment on the "grid" lines. Continuous attention to accuracy avoids compounding problems. This is especially important in multiple-row and slant-line patterns where minute joinery errors add up to become major ones in the finished work. The photos on p. 85 show how we lay up a solid-block multi-angle pattern. When you've laid up ¼ circle, check it with a square. When you've laid up ½ circle, check with a straightedge. Make a third check with a square when you've laid up ¾ circle. We usually need to touch up the last three segments slightly on the disc sander for a tight fit. Once all the segments are glued up, let the work dry overnight. We find that it's not necessary to clamp when laying up the segments, but we do clamp the layers of veneer or other hardwood laminates that may be part of the piece. We have never had a piece fly apart on the lathe.

We usually glue a piece of hardwood to the bottom of the burl. This will be turned down later to become a very thin foot at the

Photos of finished bowls: Ken Reid

Making a segmented bowl step-by-step

1. Draw desired shape full size on graph paper

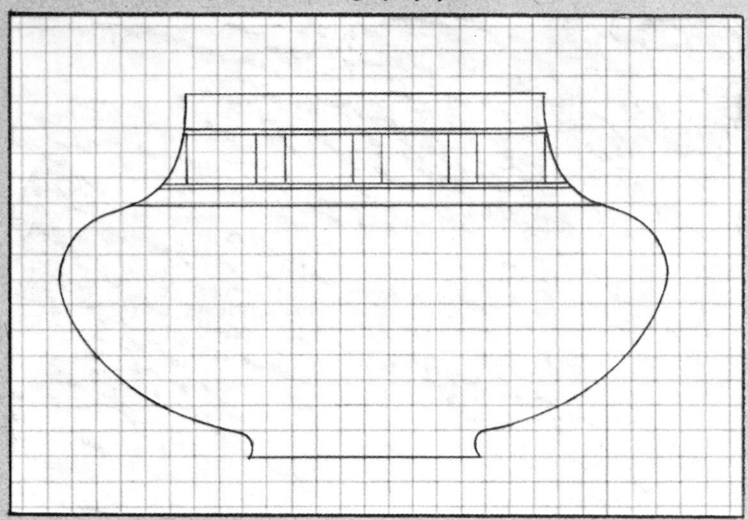

Draw plan view of each segment band:

2. Cut segments

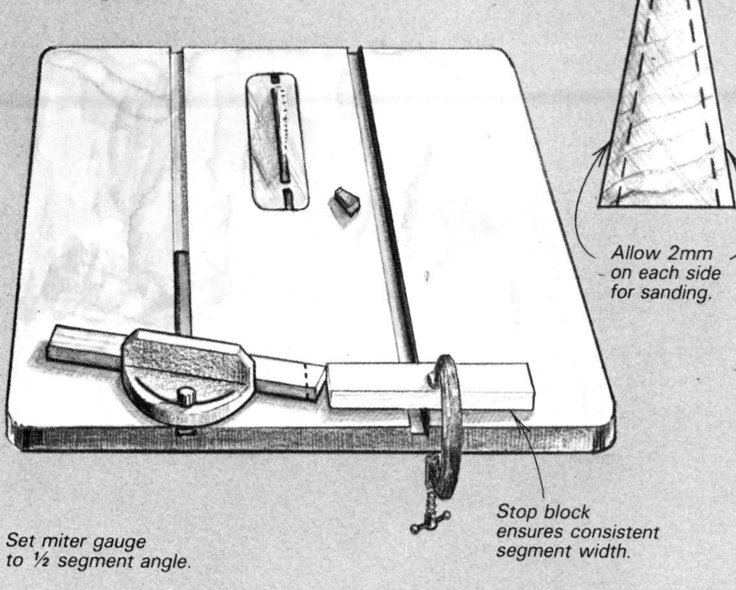

3. Sand segments to size

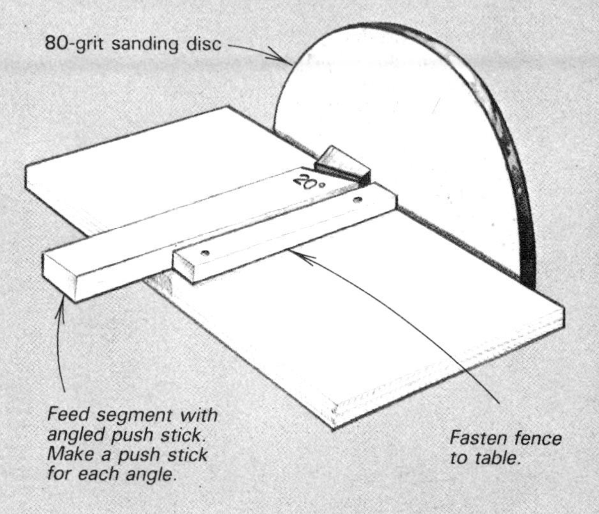

Slant-line segments

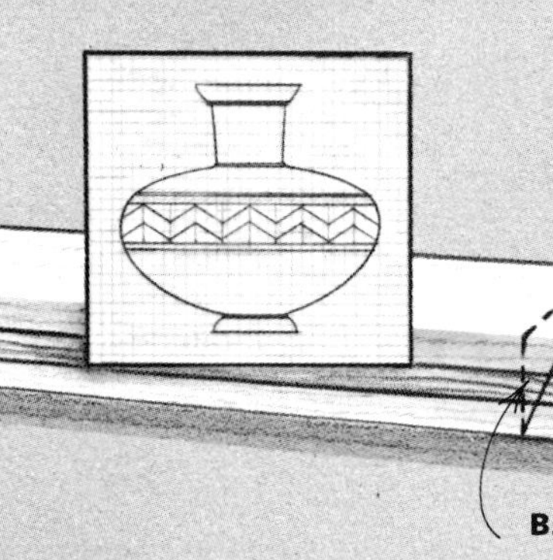

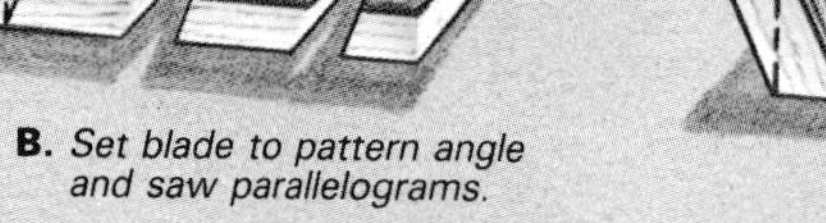

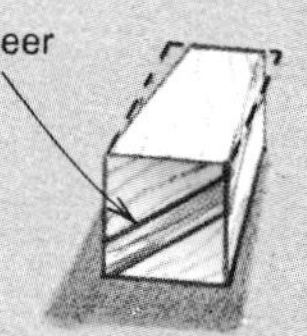

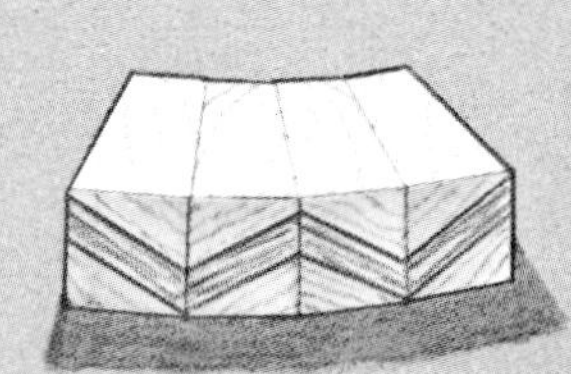

Gluing up begins with a burl-blank bandsawed round then sanded flat with a custom-made sanding disk (4). After gluing on a layer of veneer, the pre-planned segment "grid" is drawn on the blank (5) and the segments glued in place one-at-a-time. Checks for square at 90° (6), 180° and 270° reveal cumulative errors in segment size and allow corrections. The completed solid-block, multi-angle segment band (7) is set aside to dry.

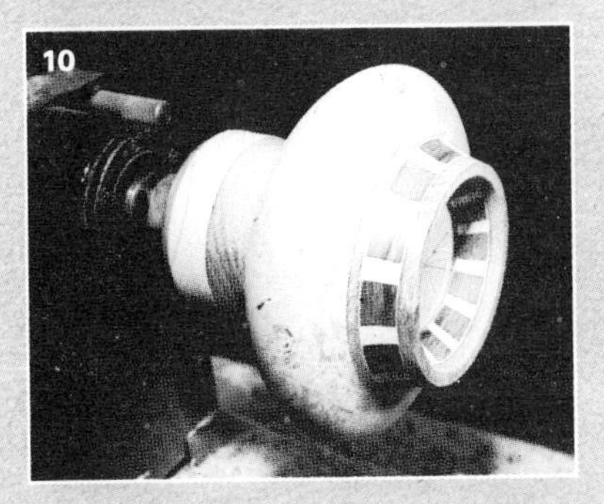

After gluing on a hardwood rim and foot, the blank is screwed to a faceplate. The center is drilled to depth with a 3/4-in. multispur bit, then the tailstock is brought up for support (8). First the outside is turned to shape with scrapers (9), then the inside is scraped, leaving the tailstock in place (10) for as long as possible to support the work. After sanding, the bowl is parted off the lathe and the bottom sanded flat before finishing with Waterlox (11).

bottom of the piece. We glue a piece of maple to the hardwood which is, in turn, screwed to a faceplate.

The blank can be quite heavy, so do as much of the turning as possible between centers. We do most of our rough turning at 1000 RPM to 1500 RPM and finish turn at 2200 RPM. First, we turn the outside with scrapers. For complex pieces we stop occasionally to check the drawing against the turning. If an interesting effect starts to happen during the turning process we don't hesitate to deviate from the drawing.

We sand the outside, first with 60-grit, then with 80-grit sandpaper before starting the inside. With a chuck in the tailstock, we bore to within 1/8 in. of the bottom of the burl using a 3/4-in. multispur bit, then turn the inside with scrapers, striving for a wall thickness of 1/8 in.

We finish-sand down to 400-grit, and part the bowl off through the layer of hardwood, leaving only about 1/8 in. still attached to the burl. We sand this flat on a piece of 120-grit sandpaper fastened to the benchtop, working through the grits down to 400-grit. Six or more coats of Waterlox finish the bowl.

The segmentation process can be as simple or as complex as you choose to make it. Since the actual lathe work is often only a small part of the entire process, it's a good idea to develop turning techniques before plunging into segmentation work. You'll find, as you experiment with our techniques, that you'll discover new effects at the lathe that you just can't predict on paper. □

Bud Latven and Addie Draper are professional woodturners in Tajique, N.M.

Drawings: Mark Kara

Laminated Bowls

Simple cuts produce complex curves

by Harry Irwin

Bowls can be turned from seasoned wood, green wood or laminated wood. The results from seasoned or unseasoned stock are similar, but the bowls produced from laminated wood are quite different. I wanted to turn some bowls but since I don't have a chain saw or a drying room, using green wood seemed beyond my capability. And I couldn't afford to purchase a large slab of seasoned hardwood either. Therefore, I turned to making bowls from glued-up wood. When I began, I did not know what form my laminations would take; the laminated bowls I had seen didn't seem especially attractive. So I decided not to look at any how-to literature and just try it on my own. The four bowls shown here are the results of my experiments.

If I could have turned bowls from solid stock I don't know if I ever would have tried this type of lamination. But now I am hooked on the idea. The field of complex laminations is new and unexplored. Haphazard gluing can be unattractive, but with some creative thinking the laminations can enhance the beauty of the wood. Glue lines may not be pretty but they are no uglier than the mortar that holds bricks together. If a bricklayer makes an elegant archway he must taper his bricks; to do so he increases the mortar-to-brick ratio. The same is true for the woodworker. If he wants to achieve a bend or design through lamination, the glue-to-wood ratio will increase. Both cases are legitimate uses of materials, and neither should be criticized for its use of adhesives. The work should be judged by the finished product.

Harry Irwin, a former carpenter, is a woodworking teacher living in Cambridge, Mass.

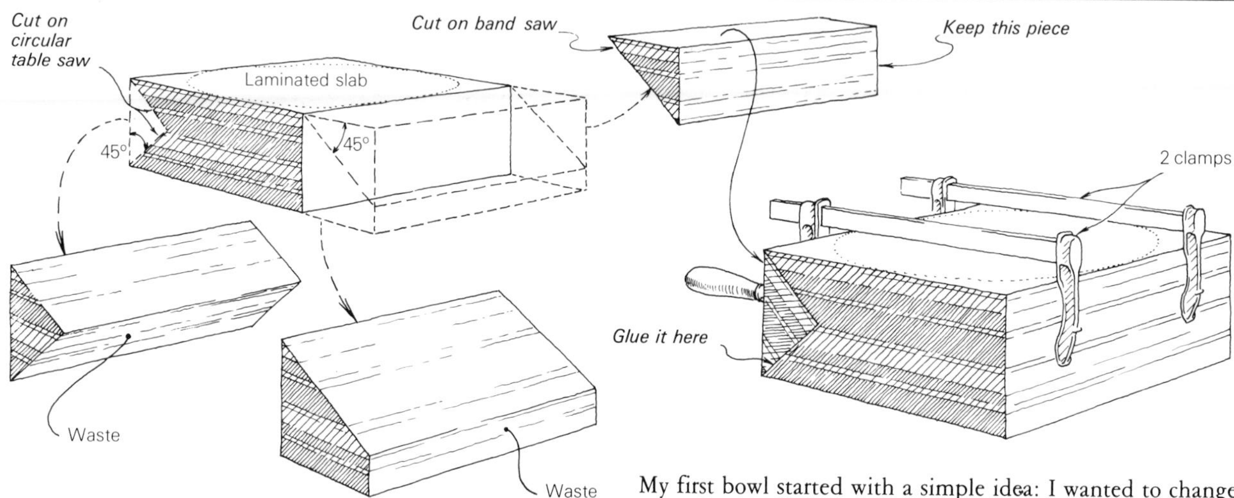

My first bowl started with a simple idea: I wanted to change the usual horizontal glue lines of a laminated bowl. I accomplished this with some risky end-grain gluing. I glued together pieces of cherry, oak, mahogany and teak. This block was clamped to the miter gauge of the table saw, for safety and accuracy. Then two careful 45° crosscuts caused a 90° chunk (a right triangular prism) to be released from the end grain, as shown in the drawing. This cut had to be carefully done because any unevenness would result in a poor and potentially weak glue line. The 90° must be precise too—it is easy to check with an accurate try square. From the other end of the block I cut the same shape, only here the grain is at a 45° angle to the hypotenuse instead of perpendicular. This cut, because of its length, must be done on the band saw. It will not be gluing surface so its flatness is of no special importance. This same end is then cut square for clamping. Because of the end grain, I sized the surfaces with a liberal coat of plastic resin glue. After it soaked and dried a little I applied some more and clamped it. The clamping procedure is very easy—two clamps will do the trick—then it's on to the lathe.

Bowl of cherry, oak, mahogany and teak, 6½-in. dia.

From *Fine Woodworking* magazine (November 1978) 13:48-49

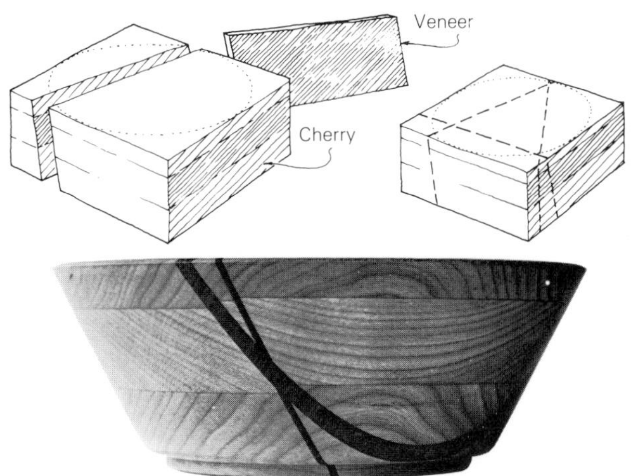

Cherry and mahogany bowl, 7⅞-in. dia.

Next, I decided to pass thin sheets of mahogany through a laminated block of cherry wood. On the table saw I set the miter gauge and the sawblade for a compound angle cut. In between the two halves I sandwiched mahogany veneer. After it dried I repeated the process two more times. In the end each piece of mahogany intersected the other two, as shown in the sketch. When the blank was turned on the lathe, the mahogany became hyperbolas. The most interesting parts proved to be the intersections of these hyperbolas.

Gluing this bowl turned out to be harder than I had expected. The angle cuts, under pressure of clamping, caused the two halves to slip apart. I solved this problem with some awkward clamping from all six sides. The unorthodox cross-grain gluing might lead to the eventual destruction of the bowl. But the thin veneers of mahogany might not have the strength to break the glue joint, just as thin layers of plywood survive their cross-grain gluing. Time will tell.

The gluing problem I encountered in the previous bowl gave me an idea for the next one. It also sent me from the table saw to the band saw. So far I was making straight cuts and the lathe was changing them into curves. This time I decided to cut a curve. I started with a block made of cherry, oak, mahogany and walnut. Through this I cut (vertically) a gentle arc. In between this arc I placed thin strips of walnut and cherry. The work was first clamped together dry, to find the gaps. Then I removed high points on the spindle sander. The final gluing was easy to do. The problem of slipping I had experienced with the previous bowl was gone, because the two arcs aligned themselves naturally. Bending wood can be difficult and time-consuming: The bending jig must be made to duplicate the curve, and steam is needed to achieve the bend. But here the bending jig is no extra work since it is also the finished bowl, while the thin laminations form easily to the arc without steam.

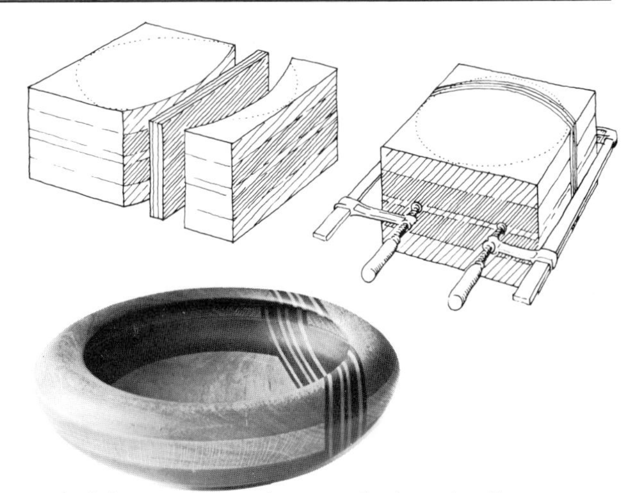

Bowl of cherry, walnut, mahogany and oak, 9¾-in. dia.

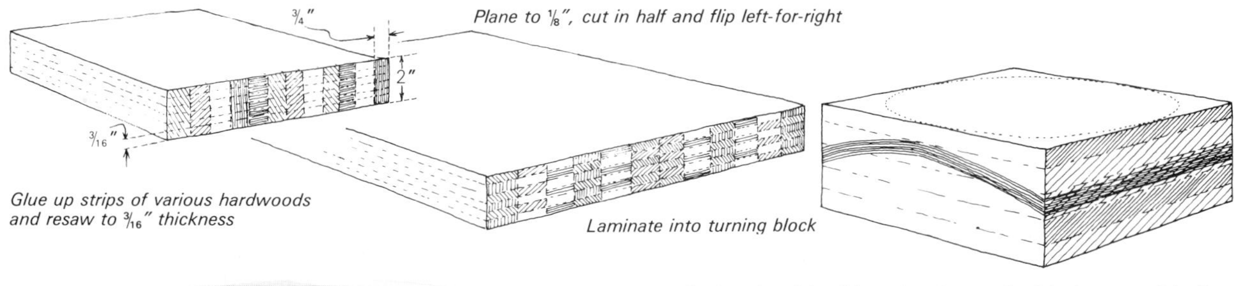

Bowl, 9¼-in. dia., of cherry, walnut, mahogany, oak, poplar, teak.

For the last bowl in this series, I started with the same block as before. But this time the arc cut through the block went along the horizontal plane. The veneer to be laminated in this space also had to be glued up. I glued strips of teak, poplar, oak, cherry and walnut, all ¾ in. by 2 in., together edge to edge. They were resawn on the band saw with the fence set at 3/16 in. A sharp blade is needed for this cut or else the blade will wander. The sheets were passed through the planer to bring their thickness down to ⅛ in. and to remove the saw marks, thus ensuring a good glue joint. They were turned left over right and at the end grain an inlay pattern appeared. Again this block was clamped dry to find the gaps that had to be sanded away. Once glued and turned, the curved laminations in the center of the bowl became a continuous wave around the bowl. □

Inlaid Turnings

Decorating with plugs

by Fran William Hall

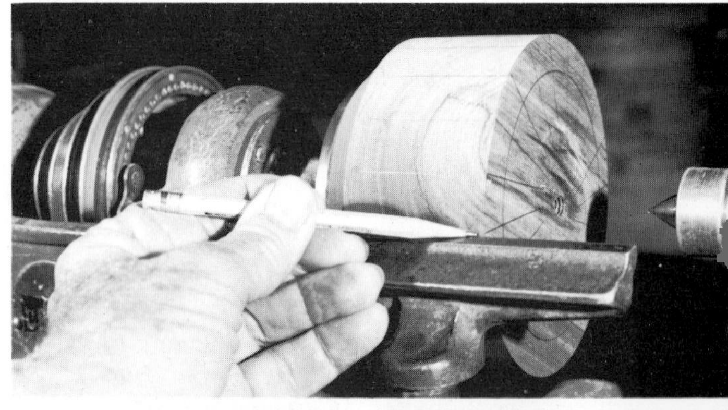

Several years ago the harvesters came through my section of southern Minnesota looking for black walnut veneer logs. They scoured the woods, buying trees and then taking only the straight bole for shipment to Germany, where, I am told, they cut veneer half again as thin as we do in America. After removing the boles, they left the rest of the trees to rot in the woods. I got permission from the farmers, most of whom I had known most of my life, to go in and remove whatever was usable. In several weeks time I had chainsawn more than 200 chunks of this lovely wood. I waxed the ends with paraffin and stored them in a shed to dry. At the end of two years I cut circle blanks from each good chunk and again waxed the perimeters to prevent checking.

I've made many bowls from this wood. The pieces that are highly figured I let do their own talking. Those that are plain I usually decorate. I first turn the bowl to a rough shape, then lay out a pattern using the indexing head of the lathe. A simple pattern, for instance, would consist of a single line of plugs all around the bowl. If the pattern is to be complicated, I must do one set of dots at a time, gluing in the plugs and then waiting for the glue to set before turning them down with a very sharp tool and the lathe running at a high speed.

To bore the plug holes, I first mark the bowl surface with a sharp punch to keep the drill from wandering. I clamp the work in a machinist's vise on the drill-press table and use multi-spur machine bits, running them at high speed. The hole must be clean and crisp its full depth, or turning will reveal ragged edges. I usually drill the holes about ¼ in. deep, but this depends on the thickness of the bowl-to-be and on what the wood looks like.

For the plugs, instead of using a plug cutter, I turn dowels. If the dowel is to be of small diameter, I use short lengths to minimize whipping, which can produce an oval section. I turn the dowel slightly oversize, then sand it with a wide piece of sandpaper (80 grit) to minimize irregularities, and gauge it to make sure its diameter is exact. Dowels must fit their holes perfectly. I never smooth-sand the dowels because a rougher surface makes a better glue joint. I use yellow glue. I cut the plugs on a bandsaw, holding them in a V-trough if they are small. I cut plugs approximately the depth of the drilled holes and, with a disc sander, put a slight bevel on each plug to facilitate entry into the hole. To prevent splitting, it is important that the plugs not be so long that they stick out any distance from the bowl. Cut them the right length and tap them in so they are level with the bowl wood or flush them off after assembly by sanding. Once again, run the lathe at high speed, use a sharp tool and take a light cut. If the plugs have been fitted properly, you should feel no crack when you rub your fingers over the finished design. □

Fran Hall is a travel lecturer whose home and home shop are in Northfield, Minn.

Inlaying decorative plugs begins with marking the bowl blank for bore centers on the lathe, using the indexing plate, top. The hole locations are center-punched and the holes are drilled with a spur bit at high speed, the blank held in a machinist's vise, center left. Center right, the holes are filled with plugs cut from lathe-turned dowels. It is important that the plugs be flush with the surface of the bowl blank, or they will split during turning. Above and right, completed bowls display the flawless surface that a sharp tool taking a light cut at high speed can produce. The bowls are of walnut, decorated with holly and blackwood.

From *Fine Woodworking* magazine (July 1981) 29:56

More Inlaid Turnings

Decorative plugs need not be made from sawn stock and inlaid symmetrically. The plugs in these walnut weed pots, left, were turned from oak branches, displaying heartwood, sapwood, ring and ray patterns in an irregular, overlapping band around the turning. The photo and method are from Dale Nish's book, ***Artistic Woodturning*** (Brigham Young University Press, 205 UPB, Provo, Utah 84602, 1980; paperback, 255 pp.; write for current price). The book is a good source of step-by-step instruction for all sorts of turning ideas, including laminated turnings, segmented forms, mosaic assemblages and turnings from wormy, rotten wood.

Decorative plugs can also be elaborate marquetry inlays. The calendar bowl, below, is the second of a series by Giles Gilson, of Schenectady, N.Y., inspired by the picture writing of primitive and modern cultures. It's based on symbols used by the American Indians to tell the twelve cycles of the moon. Gilson assembled the bowl blank in layers, beginning with its padauk base and adding alternating staved assemblies of various woods, including ash, ebony, holly and amaranth, with rings of mahogany. The central band is twelve staves of curly maple, originally 1½ in. thick and bored entirely through with an industrial hole saw to receive the inlay plug. The construction of the plugs varied. Some were cut on the bandsaw, stacking figure and background together and tilting the table to eliminate the kerf between them. Others were jigsawn after they were inserted in the stave, allowing the contrasting wood lines to be fit in kerfs that extended beyond the borders of the plug. Each plug was turned on a Unimat and measured with a micrometer to fit the hole in the stave. By the time the bowl blank was assembled, Gilson had invested three months in the project, and turning went "a little like diamond cutting."

Bob Barrett

Photography Associates/Bill Tchakirides

These bowls came not from the lathe but from single, flat boards that were bandsawn into tapered concentric rings and glued together before being shaped with abrasive discs. The bowls can be of one kind of wood or of several contrasting woods.

Un-turned Bowls

They may be round, but you don't need a lathe

by Peter Petrochko

About eight years ago, I discovered a way to make wooden bowls of many sizes and shapes, even though I didn't own a lathe or know how to use one. Playing around with a bandsaw I'd just bought, I found I could saw a flat board into tapered, concentric rings that could then be glued into a stack and shaped, using sanding discs and carving tools, into a finished bowl. I've since made hundreds of bowls this way, some in sizes and shapes that would be impossible to achieve on the lathe.

Small, shallow bowls can be made with as few as one ring; deeper ones may consist of as many as a dozen rings, all cut from the same board. For design variation, I sometimes saw the rings from a blank laminated up from several different-colored woods. Typically, my bowl blanks are boards as wide and as long as the bowl's major diameter and from ¾ in. to 2½ in. thick. The bowl's height, shape and wall thickness are governed by the initial shape cut out of the board, its thickness, the number of rings and the angle of their taper—variables that can be precisely controlled at the bandsaw. Besides it being unconventional and fun, I've found that turning a flat board into a bowl is the essence of economy.

Peter Petrochko turns boards into bowls in his Oxford, Conn., shop. Photos by Andy Badinski, except where noted. Sanding discs described in this article are sold by Sculpture Associates, 40 E. 19th St., New York, N.Y. 10003.

From *Fine Woodworking* magazine (March 1983) 39:82-84

Many bowl forms are possible—round, oval, free-form—but don't draw a shape with smaller radii than you can easily cut on your bandsaw. I use a ¼-in. wide, 6-TPI, skip-tooth blade. I mark off perpendicular reference lines, which I later use to align the sawn rings. Next I draw a guidemark completely around the top surface of the bowl blank about ¼ in. in from its outside edge. This line marks the cut for the bowl's top ring.

Before cutting the first ring (**A**), I set my bandsaw table at about 20°, an angle that I've found produces a workable taper. I start my cuts parallel to the grain because this yields a cleaner joint later on, when I glue the bandsaw kerf shut. Saw carefully and exit the blade where it entered—a tapered ring of wood is now cut loose from the blank.

To mark out the next ring, I trace the inside bottom edge of the first ring onto my blank (**B**). Before I saw, I decide on the vertical shape of my bowl, which is determined by the ring taper angle. For a shallower shape, I start with the table set at 30° or more and I increase the angle by about 5° or more per ring. Starting around 18° and increasing the taper of each ring by 3° or 4° makes a more vertical bowl. I start the second cut on the opposite side of the blank from the first so that when the bowl is glued up, the kerfs will be staggered. I saw rings in this manner until my original blank is much smaller, but not too small to be a stable bottom for the bowl. Once I've chosen the size of the base, I trace onto it the edge of the ring that will be just above it. With a gouge, I hollow out the base inside this line.

At this point, I take the pleasure of test-stacking all the rings over the base so I can see what the bowl will look like. Photo **C** shows a three-ring bowl. Before the rings can be assembled, the bandsaw kerfs must be glued. I use urea-formaldehyde glue (Weldwood's Plastic Resin) and spring clamps for each joint, with cauls to spread the pressure. Sometimes I sandwich a piece of veneer in the kerf to highlight it. I let the rings cure for a day, unclamp them, and file off any excess glue from each seam so that the rings will stack flat and tight.

I've developed a direct approach to gluing up my bowls: I pile a few hundred pounds of concrete blocks atop a bowl inverted on a firm, level surface. I sometimes use large screw clamps, similar to a veneer press, but the blocks are

Petrochko's bandsaw method lends itself to tapered shapes such as this mahogany bowl.

A. *Sawing the first ring.*

B. *Tracing the second ring.*

C. *A bowl makes its debut.*

handier, even if cruder (**D**). Be mindful of the stack's center of gravity. If it's skewed, the rings will slide out of alignment or, worse yet, the entire thing may come crashing down.

Next I'm ready to give the bowl its final shape. If you are making a round bowl and you have a lathe, you could screw it to a faceplate and finish it just like a turned bowl. Instead, I disc-sand it. I prefer industrial-grade discs made by Merit.

A scrap of rug on the bench keeps the bowl from slipping while I rough-sand the outside. Smaller bowls have to be clamped to the bench. First I shape the outside of the bowl with a 36-grit, 7-in. disc, and then I refine it with finer, 5-in. discs mounted in the drill press. I sand the inside next (**E**), starting with a 36-grit, 3-in. disc on the drill press. I wear a face shield because the coarse disc sends glue beads flying. Medium and fine sanding, done with 80-grit to 150-grit discs, take much less time. After sanding, I finish the bowls with either mineral oil or 8 to 16 coats of Behlen's Salad Bowl Finish. When the finish has dried, I sand with fine wet-or-dry sandpaper, using vegetable oil as a lubricant, followed by fine steel wool. For the final sheen, I buff my bowls with a pad charged with polishing compound. □

D. *Clamping the bowl.*

E. *Sanding the inside.*

Bandsawn Baskets

Spiral your way to a collapsible container

by Max Kline

The idea for these collapsible baskets came from those fold-up drinking cups that were popular as novelties years ago. You simply bandsaw a continuous, spiraling kerf into a flat board at a slight angle, so the segments wedge against each other when the basket is opened. These baskets are useful for holding all sorts of things, from fruit to floral arrangements. When collapsed, they make nice tabletop trivets.

I make the oblong baskets from a ¾-in. thick board, 12½ in. by 6½ in. You could make larger baskets, but I don't recommend smaller ones because accurate sawing becomes difficult. I've made round baskets using the same technique, but I don't find them as attractive. Any wood species will do, although it should be knot-free, not too brittle, and resistant to checking. Very hard, light-colored woods such as maple and ash will sometimes burn when cut, particularly if the blade is dull. This burning is practically impossible to sand off.

To make a basket, scale up the pattern shown in the drawing and attach it to your board with rubber cement or double-faced tape. A new pattern is needed for each basket, so I photocopy them a few at a time to keep on hand.

As the drawing indicates, cut the handle with the bandsaw table set at 90°. I use a ¼-in., 6-TPI, skip-tooth blade. After the handle has been cut and removed, tilt the saw table to about 6°—an angle that seems to produce the best bevel to wedge the segments together when the basket is opened. Increasing the angle makes a shallower basket, but if you decrease it too much, the segments will drop through without wedging and you'll have a wooden "Slinky" toy. Thicker wood, say, 1-in., can be sawn at about 5°. As you saw around the pattern lines, the board will become harder to

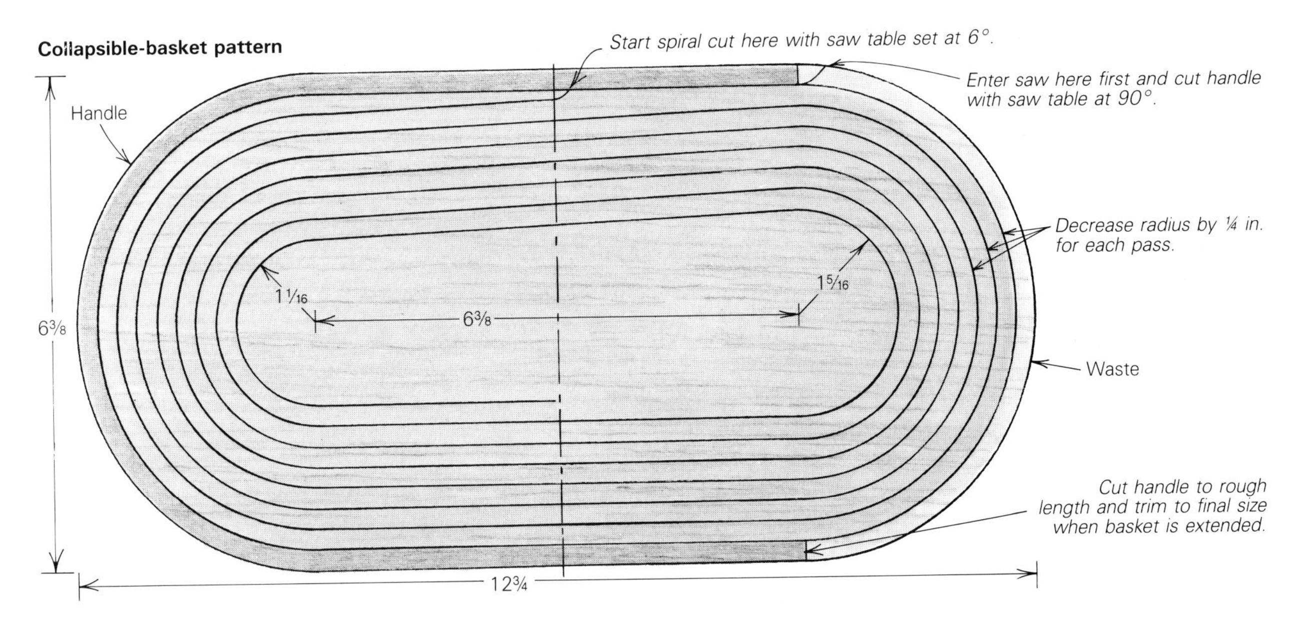

With a bandsaw set to cut a 6° spiraling bevel, Kline saws flat boards into oblong and round baskets. Below right is a hackberry basket made with the pattern shown in the drawing above. The round mahogany basket, below, is shown collapsed and inverted. Kline screws crosswise cleats to the basket bottoms to serve as both feet and pads for the handles, which hold the baskets open.

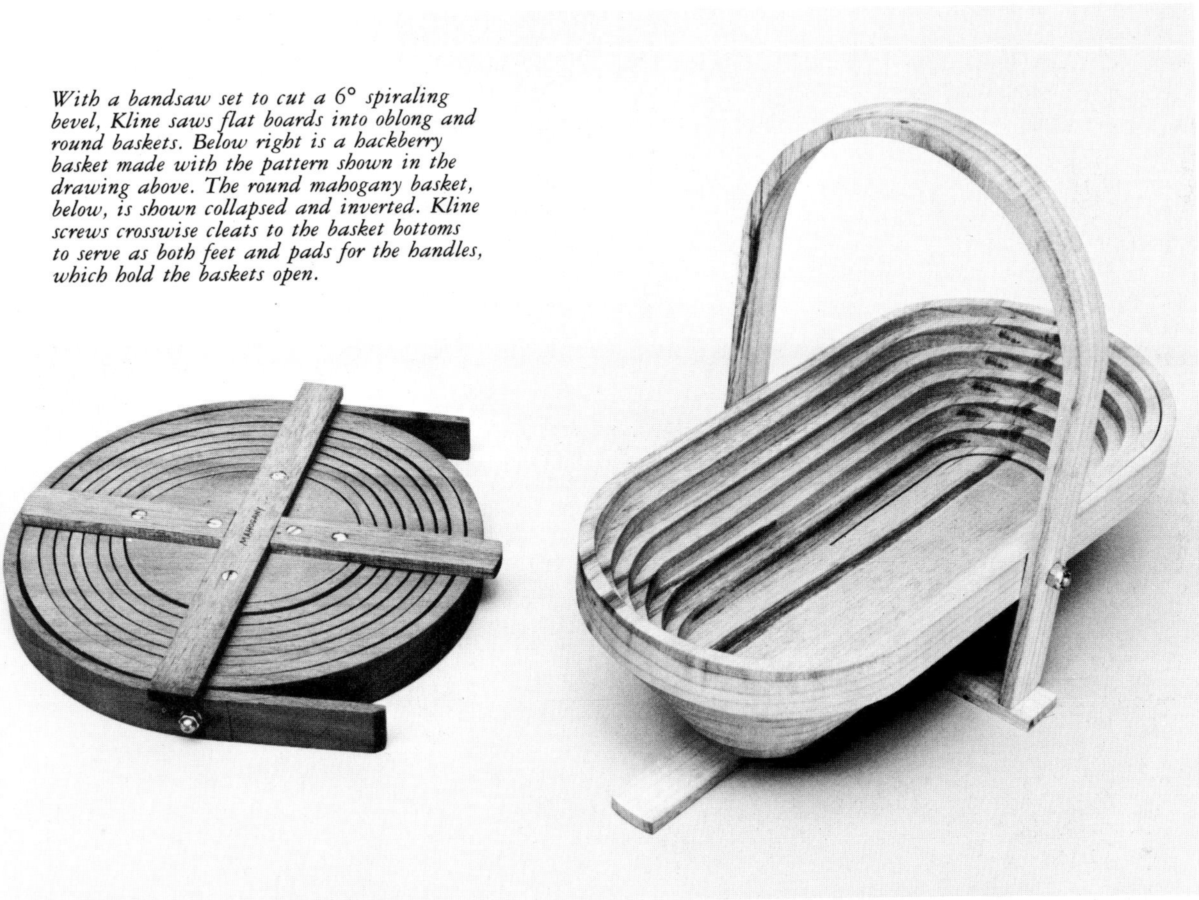

control because the spiraling cut makes the blank springy. To maintain an accurate cut, grip the stock tightly at the sides and squeeze the segments together. When you reach the end of the pattern, back the blade out.

Next, glue back in place the free end where the saw entered. I sand the handle and the outside contours with a 1-in. belt sander. If you follow your pattern accurately, you shouldn't have to sand the inside of the basket. Attach the handle with flat-head stove bolts countersunk on the basket's inside and fastened with brass or chrome cap nuts. My pattern produces an extra-long handle which must be trimmed and positioned so that it holds the basket open at its full height but stores snugly when the basket is collapsed. Screw or glue the two lengthwise cleats and one cross-cleat to the basket bottom, and finish as you desire—I use Watco oil, followed by wax. □

Max Kline is a retired chemist. He lives in Saluda, N.C.

From *Fine Woodworking* magazine (March 1983) 39:84-85

Turning a Lidded Box

A centerwork project

by Richard Raffan

Lidded boxes may seem complicated, but the steps involved are really quite simple. Boxes demand more precise tool control than do bowls, and care, attention to detail, and a few tips on how to overcome all the little problems usually encountered make them readily achievable turning projects. Craftsmanship has less to do with the conception and birth of an object than with knowing when to be careful and what to do when things go wrong.

I've made boxes as large as 10 in. in diameter and 6 in. deep. These were turned on a faceplate with the grain running across the lid and base, but warping always spoiled the lid fit when the grain was aligned this way. Today, I make all my boxes with the grain running through from top to bottom. What little warping does occur is not much of a problem on a small box because the lid can be made thin enough to flex slightly without being too fragile. But I find warping is still a major problem in boxes over 3 in. in diameter, even with well-seasoned wood.

For turning boxes, I prefer what's known as a spigot chuck (available from Cryder Creek, Box 19, Whitesville, N.Y. 14897). This chuck grips a short tenon or flange turned on the end of the wood. A 3-jaw chuck may also be used for turning lidded boxes. I do not recommend screw chucks for boxes because they don't

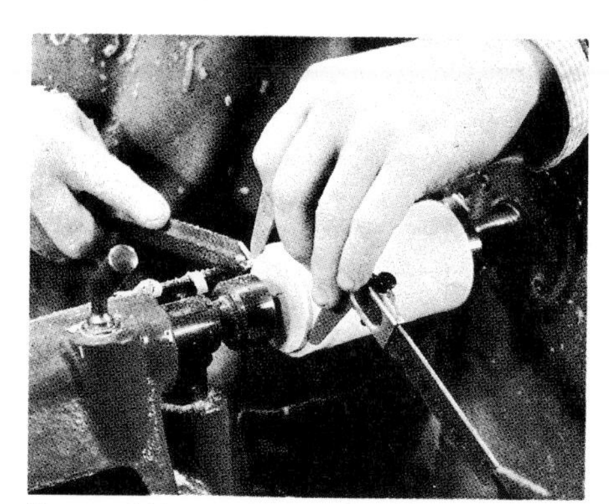

Turn a ⅛-in. tenon or flange on each end of the cylinder to fit the spigot chuck. Hold the parting tool in one hand and the calipers in the other. Stop cutting when the calipers slip over the tenon.

Tenon for 3-jaw chuck

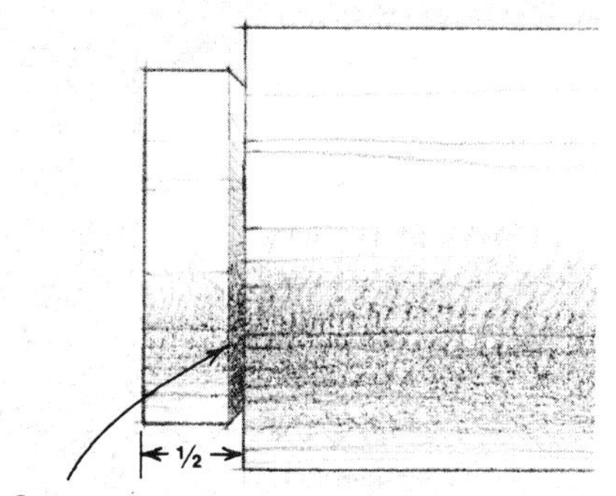

Cut groove to prevent chuck from tearing end grain.

A second shearing cut with the skew chisel trues up the rim of the lid. Tilt the short point of the skew away from the wood to avoid a catch.

From *Fine Woodworking* magazine (November 1985) 55:50-53

grip well on end grain unless the thread penetrates the wood an inch or more. This wastes wood and develops leverage problems that do not arise when working closer to the headstock. Neither do I recommend expanding collet chucks for boxes. As they expand into a recess they act like mini log splitters and tend to weaken the wood. If a tool should catch, especially at the point farthest away from the chuck, it will likely lever the blank away from the chuck and split the wood.

To start, turn a cylinder between centers with the lathe running no faster than 1200 RPM to 1500 RPM. A 2-in.-dia. cylinder 4 in. long is a good size. With a parting tool, turn a tenon on each end to fit your chuck. The size of the tenon will depend on the type of chuck. A spigot chuck will grip a ⅛-in.-long tenon. A 3-jaw chuck needs a ½-in.-long tenon with a groove cut in the corner where it protrudes from the main cylinder. This will prevent end grain being pulled by the jaws as they clamp in to grip.

Mark off the lid and bandsaw the cylinder in two, giving you separate blanks for the lid and base. Mount the lid blank in the chuck and true it by making shearing cuts along the cylinder and across the end grain with a small skew chisel. Take the opportunity to practice tool technique. Choose the technique you find most difficult and practice now, while a catch is not too disastrous.

Once you have trued the end grain, take a final cut ⅜ in. in from the rim before hollowing the interior, as shown in the photo on the facing page. Undercut this surface slightly so it fits flush with the shoulder against which it will eventually rest. With very hard woods such as cocobolo, African blackwood or Mulga, the cleanest surface will probably come from a very delicate scrape cut.

Next, I rough out the domed inside of the lid with a ¼-in. or ½-in. shallow-flute fingernail gouge. I use an old trade technique, cutting away from the center to 2 o'clock, as shown in the drawing at right. Position the tool rest so that the gouge point is at the center of the stock and begin the cut with the gouge on its side, flute facing away from you. Push the tool in at the center about ⅛ in., then pull the handle toward you and simultaneously rotate the tool clockwise to keep the bevel rubbing and the edge cutting. (The tool really does cut upside down on the "wrong" side of center.) Hollow the lid with a series of cuts, starting at the center and working outward with each successive cut until the walls are about ⅜ in. thick. Finish shaping the inside with a heavy roundnose scraper, taking light cuts.

You must now consider how the lid fits and how the desired suction fit between lid and base (see box at right) can be achieved. Two points here: first, the suction comes from the two cylinders sliding apart. The finished flanges on the lid and base must not taper. If they do, you'll end up with a lid that fits tightly, but you'll never enjoy the gentle resistance of the suction as you remove it. Secondly, all parts of the lid that will contact the base must be turned as accurately and cleanly as possible so that they fit true on similarly turned parts on the base. Sanding must be kept to a minimum to avoid eccentricity as softer grain is worn away. Cut the fitting parts well enough so that only a quick dab with 180-grit sandpaper is required for a smooth surface.

With a square-end scraper, rough out the flange leaving about $\frac{1}{32}$ in. more than your finished surface. Take a final cut with the scraper to finish the flange. Be sure to grind a sharp left corner on the scraper edge. Check the flange with inside calipers to ensure that you have a true cylinder (no taper). This is the first part of the perfect fit.

During this stage your tool may catch and knock the blank off

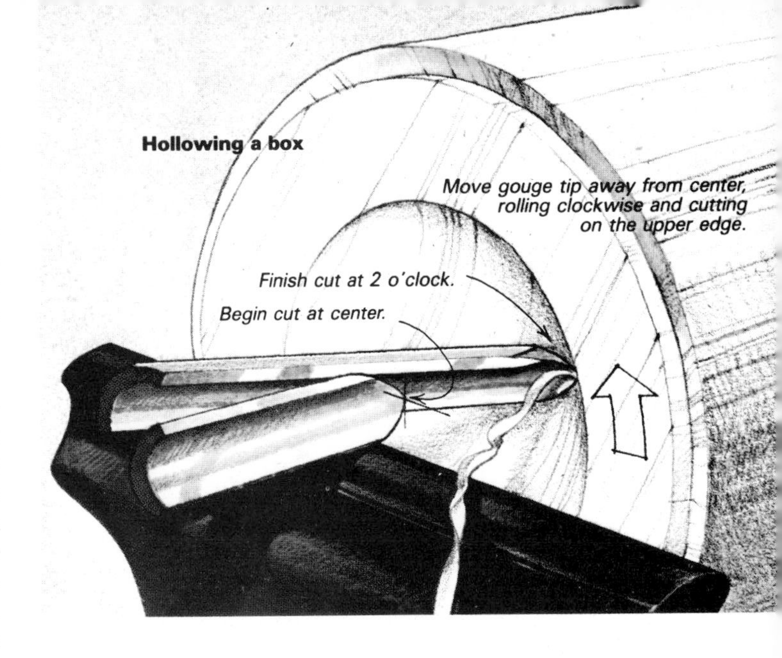

About box design

I like box lids to fit so they pull off easily against the resistance of a slight vacuum and fit against a cushion of air created as they slide over the base. I sometimes test the fit by lifting the box by its lid. It should take about two seconds for the base to slide off a perfectly fitted lid. I like the interior of the box to be a different shape from the exterior, so that it might surprise the inquisitive. The inside contour doesn't need to follow the outside.

To disguise any movement in the wood, I detail the line where the lid and base meet with a groove or a bead. A smooth join on a freshly completed box will be hard to detect, but later (usually only minutes), the slightest eccentricity or warping will leave one edge jutting over the other to mar the surface for a caressing hand. Detailing the join eliminates this problem.

The line of the join affects the visual balance of the box. Mostly, I prefer to locate it between one-third and one-half of the way from either the top or bottom, but if I don't care for its position once I've cut it, I'll add other bands or grooves to balance the form. ***—R.R.***

Photo this page: Richard Brecknock

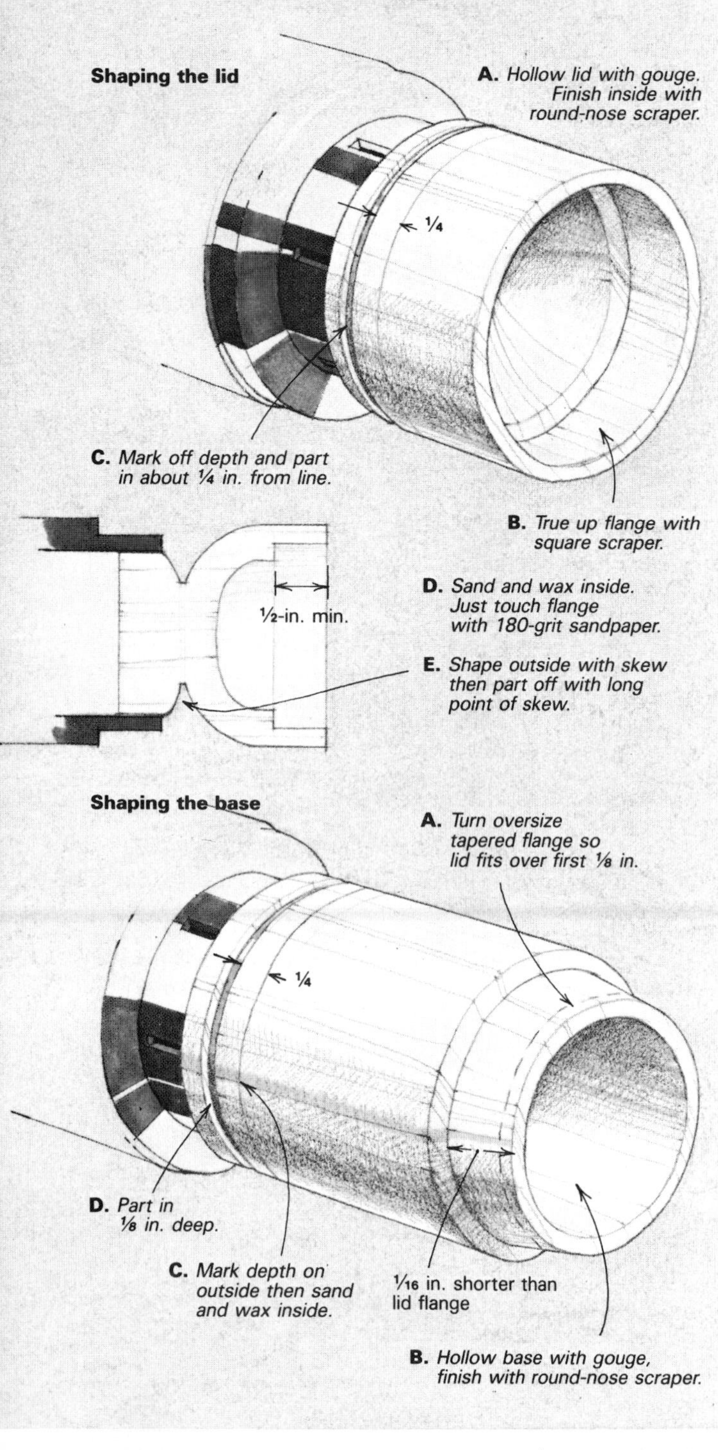

After sanding and waxing the inside, trim the flange to fit the lid. Use the long point of the skew as a scraper.

center. Don't worry. Remount it and true the inside dome of the lid, leaving the flange and rim until last. If you've cut the inside and still need to true the rim, don't use a shear cut because the grain will split away down the flange. Use a delicate scrape on the end grain.

Measure the depth of the lid and mark a pencil line on the outside. Sand and finish the domed inside of the lid. Be careful not to touch the flange, which should require only a dab of 180-grit sandpaper. I finish with soft beeswax.

To define the top of the lid, part in about ⅛ in. from the line on the headstock side. Rough out the exterior of the lid with a skew, then part off the lid with the point of the skew. You'll finish turning the lid later, when it's mounted on the base, but cut as much as possible now, while the blank is held firmly in a chuck.

Mount the base blank and true it with the skew. To rough-fit the lid, cut a tapered flange so that the lid fits just over the end. This is surprisingly easy to do by eye, but if you make the end too small, just extend the flange farther back into the blank. As the base revolves, fit the lid and apply just enough pressure for the lid to leave a burnish line. This line gives you the final flange diameter. Don't cut the rest of the flange to size yet. If you hollow the base first, you can afford a massive catch and get away with it. If you finish the lid fitting and then have a catch, you'll probably fail to get the base running true and will have to start over again.

Hollow the base with a ½-in. gouge followed by a roundnose scraper. Measure the depth and mark this on the cylinder, then sand and wax the inside. To mark off the bottom, part in ⅛ in. from the line on the headstock side. This gives you a ⅛ in. thickness for the base. (Make it ¼ in. if you're really nervous.) Don't part in deeper than ⅛ in. at this stage. You need to know where the bottom is when you finish turning the exterior, but you still need the support of the wood running into the chuck.

Using the long point of the skew as a scraper, cut away the flange taper so that the lid fits tightly. If at this stage you discover the flange slightly off center, it doesn't matter. Turn it true. If you've overcut it, you can cut the flange shoulder farther back into the base and, if necessary, cut some off the rim. Cut the flange about 1/16 in. shorter than that in the lid, and cut the shoulder at the bottom of the base flange cleanly. Ideally, the fit between lid and base will be tight enough to prevent the lid from spinning on the base when you remount the assembled box for final shaping of the lid.

If you have a good suction fit, but not enough friction to prevent the lid from slipping and spinning on the base, try this: remove the lid and hold a lump of soft wax (beeswax is ideal) against the revolving flange so that a ring of viscous wax develops. Stop the lathe and mount the lid before the wax solidifies. You have only a few seconds to push the lid on but once there, the cooled wax will hold it fast unless you cause the lid to turn slower than the base by cutting or sanding too hard, in which case friction quickly melts the wax.

Turn the outside with a skew chisel. Depending on your skill and audaciousness at this stage, you can turn a delicate finial on the lid. This isn't difficult as long as you put no pressure against the axis. Arc the point of the skew down into the wood by pivoting the skew on the rest for maximum control. Don't merely push the skew forward into the wood.

Sand and finish the outside before fine fitting the lid. This is the stage that makes or breaks the quality of a lidded box—getting that suction fit just right. With practice and experience it can be done within a minute. Otherwise it takes time and patience.

Drawings: Joel Katzowitz

Fit the lid on the base and finish shaping the box with the skew chisel.

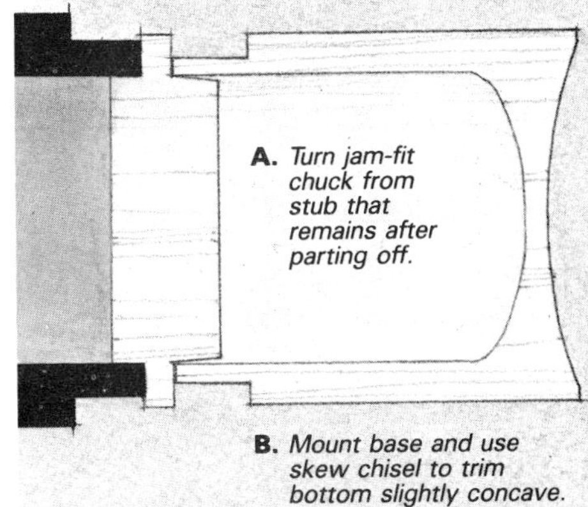

Proceed with caution. Too much enthusiasm at this stage and you could overcut and the lid will be loose. The best fit will come from a tool-cut surface with a minimum of sanding. I use the long point of my skew chisel as a scraper. This gives maximum control with minimum risk. After each delicate cut I can stop the lathe, try on the lid, and test the fit. Once it pulls off with reasonable ease, I sand the flange—a dab of 180-grit sandpaper is sufficient—and wax.

Once the lid fits satisfactorily, part off the base. Be careful to catch the box, not hold it, or the wood still attached to the chuck will spin a hole in the bottom.

On the stub that remains in the spigot chuck, turn a tapered jam-fit chuck, as shown in the drawing. Mount the base and true up the bottom with a skew chisel. I always turn the base slightly concave so that the box sits flat. I usually chamfer the corner between side and bottom using the long point of the skew. A sharp corner could easily be damaged or chipped. □

Richard Raffan, a professional turner, lives in Mittagong, N.S.W., Australia. His book Turning Wood with Richard Raffan *and the companion video of the same name are both available, separately or as a set, from The Taunton Press.*

Poured pewter inlay

by William Vick

I decorate my turned boxes with pewter inlays. Pewter, an alloy of tin, antimony, copper, and sometimes bismuth or lead, has a low melting point (420°F) and is easily poured into kerfs cut by lathe tools. One source for pewter is T.B. Hagstoz, 709 Sanson St., Philadelphia, Pa. 19106.

To inlay a flat lid, rough the outside to the final shape, then use a parting tool to cut kerfs at the desired locations. The kerfs should be at least ⅛ in. deep and slightly undercut. The undercut serves to anchor the pewter.

To inlay a band around the circumference, turn the area above the band close to the finished box diameter. Form the groove by cutting in at an angle with a parting tool, leaving a dam to contain the molten pewter.

To melt and pour the pewter you'll need a pouring ladle with a wooden or plastic handle (a ladle with a wooden handle and small spout is available from Dixie Gun Works, Union City, Tenn. 38261) and a propane torch. The box must be on a perfectly level, non-flammable surface. In a well-ventilated area, away from combustibles, put a small piece of pewter in the ladle and melt it by heating the base of the ladle with the torch. Once it melts, continue heating for about 30 seconds more. The metal must be hot enough to flow completely around the inlay cavity. Pour quickly and evenly. If the metal hardens before the cavity is completely filled, you'll end up with defects in the finished inlay.

When the pewter has cooled, mount the piece on the lathe and take light cuts with a sharp tool to trim the piece to final shape. Cut the pewter and the wood together. Because pewter is so soft, the cutting edge will not dull quickly. □

William Vick teaches woodworking at Mills Godwin High School in Richmond, Va.

Pewter bands were poured in place.

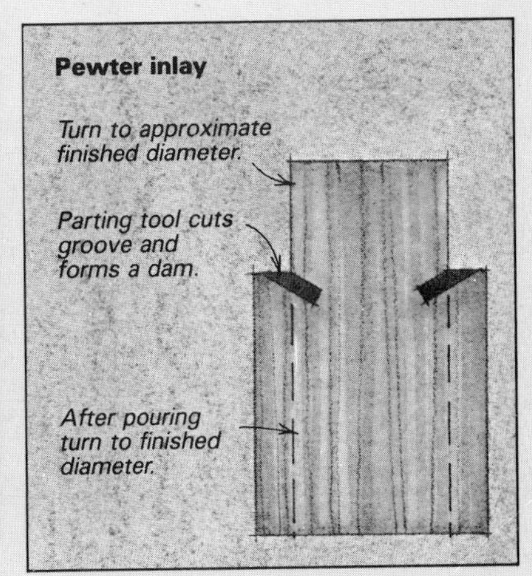

Small Turned Boxes

Grain direction determines technique

by Wendell Smith

Walnut box, 6-in. diameter.

Small wooden boxes offer the woodturner an excellent opportunity to try his hand at design. I would like to describe some techniques I use to make boxes for storing small items such as jewelry. One approach to the design of such turnings that I particularly like is to inlay circular veneers into box lids. The availability of many figured veneers allows an almost unlimited number of attractive wood combinations. Some wood/inlay combinations I enjoy are walnut/thuya burl, padauk/ebony, maple/ebony, cherry/tulipwood, mahogany/madrone burl, mahogany/amboyna burl, walnut/ *Dalbergia sp.* (rosewood, cocobolo, kingwood, etc.).

Depending upon whether the grain is to run horizontally or vertically in the completed box, the sequence of operations in box turning is somewhat different. I've divided the discussion according to this difference.

Grain is horizontal

Turning boxes of horizontal grain involves primarily faceplate techniques. Useful lumber thicknesses are 2 in. and 3 in. With 2-in. stock, I generally use the full thickness for the base, and resaw another piece from an adjacent section of the same board for the lid. This point is quite important, because differences in color and texture between boards can be very apparent if one is used for a base and another for a lid. A 3-in. blank may be resawn into 2-in. and 1-in. pieces for the base and lid, thus permitting the figure to carry through from top to bottom. Resawn lid blanks should be given ample time to return to equilibrium with the ambient humidity prior to turning. Then glue round waste blocks of 3/4-in. plywood with paper interleaving to the bottom of the base and the top of the lid. Polyvinyl acetate (white) glue and brown wrapping paper serve the purpose satisfactorily. The faceplate, of course, is screwed to the waste block.

I turn and finish-sand the inside of the lid first. The lid is then removed from the lathe and the glue block split off. The base of the box is placed on the lathe, and a shallow recess turned to accept the lid with a snug fit. The base is now capped with the lid, and the two held together with a flat center in the ball-bearing tailstock. The support of the tailstock eliminates the necessity for a tight jam fit. Unless a jam fit is very tight, slippage may occur when turning the base and the lid together, which invariably burnishes the wood. And a very tight fit may require prying to remove the lid, which can easily damage the box. Before capping the base with the lid, the top edge of the base is sanded. Thus, those surfaces on the base and lid that contact each other are now complete, and so is the inside of the lid. With this arrangement, the side of the box and lid may be worked together, and then finish-sanded. Before removing the tailstock, the lid is "clamped" to the base by wrapping the joint with masking tape. I use 1-1/2-in. tape, which is amply strong.

The top of the lid is now finished as desired. One precaution: If a knob is desired, it is usually better to add a separately turned one rather than to turn the lid and knob from the

---

Wendell Smith, of Fairport, N.Y., is a chemist in the Kodak Research Labs in Rochester, N.Y. His turnings, which he sells through art galleries, have been exhibited in several juried craft shows. Smith uses a Rockwell-Delta lathe (company name is now just Delta) and Sorby gouges.

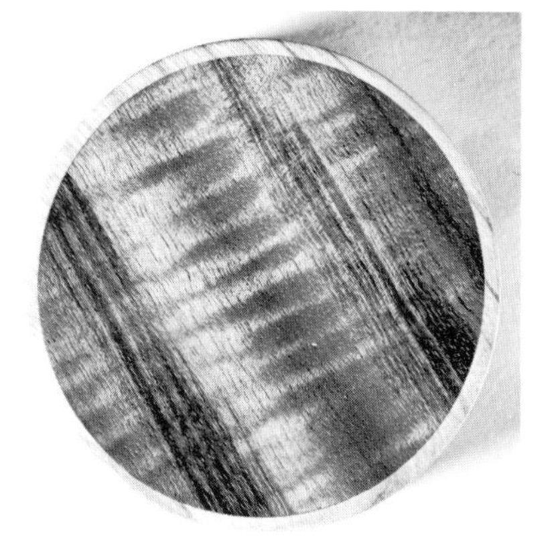

Grain is horizontal in 3-1/2-in. diameter cherry box, left, with koa veneer inlaid into its lid, center. Box at right is goncalo alves, 4-1/4 in. in diameter. Although both boxes are plain cylinders, subtle changes in shape make each distinct.

From *Fine Woodworking* magazine (Winter 1977) 9:72-74

same piece. Because a large amount of wood must be removed from the top of the lid to leave a knob, re-equilibration of the lid with the ambient humidity may lead to cupping.

Upon completion of the top of the lid, the masking tape and lid may be removed and the interior of the box finished. Boxes in which the grain runs horizontally rather than vertically are more prone to develop sticking lids with humidity changes. Consequently, the interior of the box should be sanded until the lid fits loosely. For this type of box, a loosely fitting lid is a properly fitting lid.

There is a trick to removing waste plywood blocks that prevents damage to the bottom of the finished turning. The splitting wedge (an old plane blade) should be inserted between the two layers of the waste block closest to the paper interleaving—not between the waste block and the paper. This removes about 80% of the plywood. The remainder, still glued to the paper, has no strength and is easily pried off with a chisel. Scraping and sanding complete the job.

To inlay a veneer into a box lid, the first step is to soak, press and dry the veneer. Once dry, the veneer should be kept under moderate pressure to keep it flat until ready for use. It is then clamped with a circular wooden block against a wooden faceplate previously flattened in the lathe. The wooden block is held in place with a ring center in the ball-bearing tailstock. The ring center indexes the block for later re-centering at the gluing stage. The block should be about 1/8 in. smaller in diameter than the desired circle of veneer.

A tool rest is now brought as close as possible to the veneer and block. With the lathe running at a slow speed, the veneer is cut using a skew or diamond-point chisel. Using this method I have cut circles with undamaged edges even from recalcitrant veneers such as burls.

To prepare the box lid for the veneer, I first turn a slightly undersize recess. A "straight-across" scraper is used to flatten the bottom of the recess. The diameter of the recess is carefully enlarged by using a parting tool to pare off a small amount at a time. This means frequently stopping the lathe to test the fit, a tedious but necessary process. The fit should be snug—anything less results in an unsightly glue line. I remove the tightly fitting veneer for final gluing with the shop vacuum cleaner, using the last gentle gasp of suction after the vacuum has been turned off.

To attach the veneer, I use a polyvinyl acetate (white) or an aliphatic resin (yellow) glue, both of which are water-based. Put the glue on the box lid—not on the veneer. A water-based glue applied to the veneer will cause the veneer to expand rapidly and it will be difficult to fit it into the recess. Finally, the inlay is clamped in place in the lathe using the same circular wooden block (previously indexed with the ring center for re-centering) as was used in cutting the veneer, with waxed paper inserted between the veneer and block. After overnight drying in the lathe, the lid may be finished.

Grain is vertical

I generally turn vertical-grain boxes (such as a ring box) from 2x2 spindle-turning stock. Because the lid is separated from the base on the lathe, allowance must be made for some waste.

The blank is first placed in the lathe between centers and turned to a cylinder. A shoulder is then turned at one end to fit tightly into a homemade chuck, as shown in the diagram,

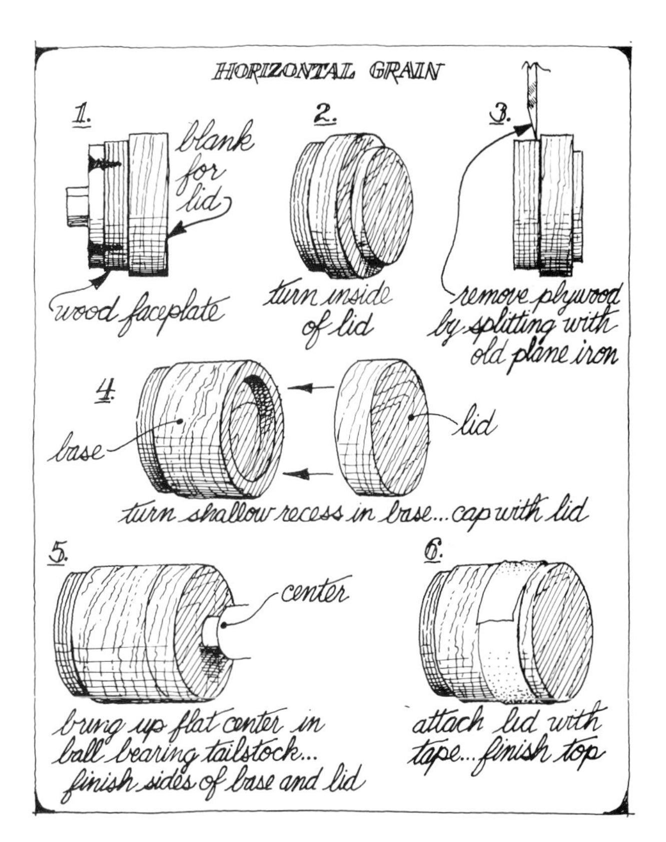

Horizontal-grain box in cherry, 10 in. diameter, 2-1/4 in. high, shows design possibilities permitted by the method diagrammed above. Lid is mostly finished with tailstock for support, then taped in place for final shaping of knob.

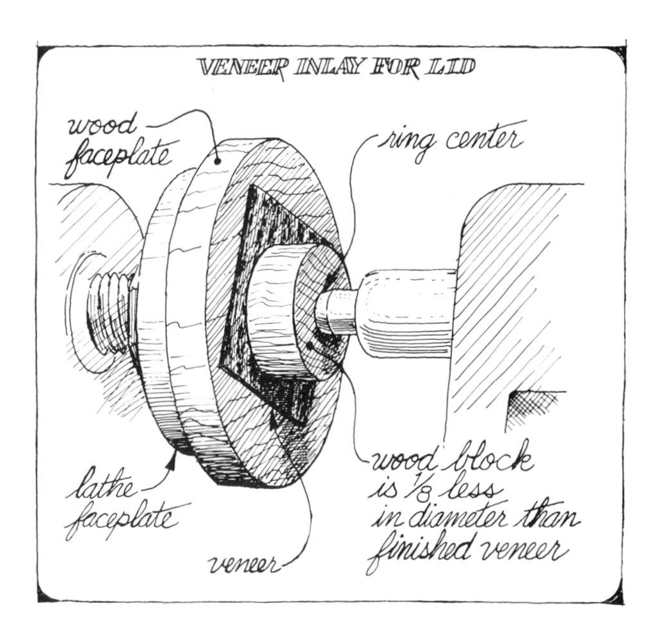

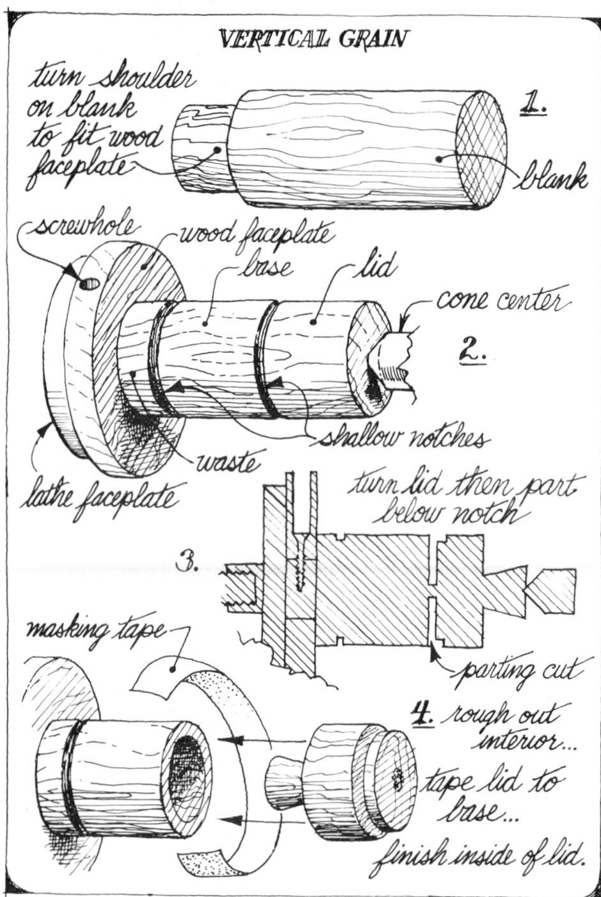

and securely fastened with two countersunk screws.

The lid can be prepared in two ways. If the wood is figured, it is desirable to have the figure carry through from base to top in the completed box, as shown in the photograph. Consequently, when the lid is parted from the base, it must go back on the same way it came off. If the wood has little or no figure, it is somewhat easier to prepare the lid in the reverse sense; that is, after parting from the base, the lid is turned around and refitted. Since the method for figured wood is more involved, I'll describe it in detail.

The cylindrical blank, fastened to the chuck, is mounted in the lathe with a cone tailstock center supporting the free end. After truing so that the cylinder runs smoothly, a shallow notch is cut with the parting tool where the lid will meet the top of the base. A second shallow notch is cut at the bottom of the base, no closer than about 1/2 in. from the chuck. These notches assist in visualizing the relative proportions of the base and lid and other aspects of the design of the box. In estimating the height of the lid, about 3/16 in. extra should be allowed at the tailstock end for later removal of the cone center mark. Shrinkage and expansion are not much of a problem with vertical-grain boxes, so a knob may be turned as an integral part of the top.

The knob and top of the lid are first roughly turned to the desired shape. Then the lid is parted from the base, the parting tool entering the cylinder about 1/8 in. below the notch previously cut between base and lid. The interior of the base is next partially roughed out to accept the inverted lid, which is fastened to the base by wrapping the two with masking tape. This automatically centers the lid, and its inside, including the shoulder, is now easily finished. A slicing cut with the skew is ideal for finishing the bottom of the lid, as this exerts little force and leaves a surface that requires a minimum of sanding. After sanding the bottom of the lid and shoulder, the masking tape and lid are removed.

The turning of the remainder of the box is similar to the method for horizontal-grain boxes. The interior of the base is first enlarged so the lid will fit snugly. The tailstock is brought up, and the side of the box and lid are worked together. The joint is then taped, the tailstock removed, and the handle and top of the lid completed. One must take very thin slicing cuts at the end of the handle to avoid breaking it. Finally, the lid is removed and the interior of the box is completed.

I remove the box with a parting cut that goes nearly to the center of the cylinder. To prevent fibers from being torn from the center of the base, I terminate the parting cut when about 1/4 in. of wood remains. Parting is completed with a coping saw, leaving the 1/4-in. nubbin of wood on the box. The box is then turned around and snugly fitted to the approximately 1/2 in. of waste wood still protruding from the chuck, by turning the waste wood to size. The box is supported with the tailstock cone center touching the nubbin, and all but the very center of the base can be cleaned up with a slicing cut of the skew, followed by light sanding. The nubbin is then removed with a jeweler's saw and any remaining waste sliced off with a chisel. Hand sanding completes the base. □

Homemade chuck holds cylinder with vertical grain, photo at top, for turning sequence shown in the drawing. At left is soft maple box, 1-7/8 in. in diameter and 5 in. high, with separately turned knob of walnut. Myrtlewood box, right, is 2 in. in diameter and 3-3/4 in. high.

Chatterwork

A risky path to a faceted finish

by Stephen Paulsen

For six years I've been perfecting a technique for decorating small spindle turnings with three-dimensional surface texture. I call it chatterwork. It's an efficient way of producing mandala-like patterns that resemble those cut on a 19th-century Holtzapffel lathe, or David Pye's meticulously turned containers decorated with patterns made by ornamental turning attachments. Chatterwork can be done on any lathe, and takes minutes instead of hours, although it looks as if much more time were involved. Since this technique also cuts sanding time, it's economically feasible for production turning.

While I can imagine many applications for chatterwork, from drawer pulls to dowel caps and decorative inlays, I use it mostly on the stoppers for my glass-lined wooden scent bottles. Hard, heavy, dense woods—ebony, brazilwood, rosewood and African blackwood—are best for chatterwork because they hold the sharpest detail.

You may already have inadvertently

Chatterwork graces the lids and insides of Paulsen's tiny boxes, and the spired stoppers of his glass-lined wooden scent bottles.

Photos this page: White Light

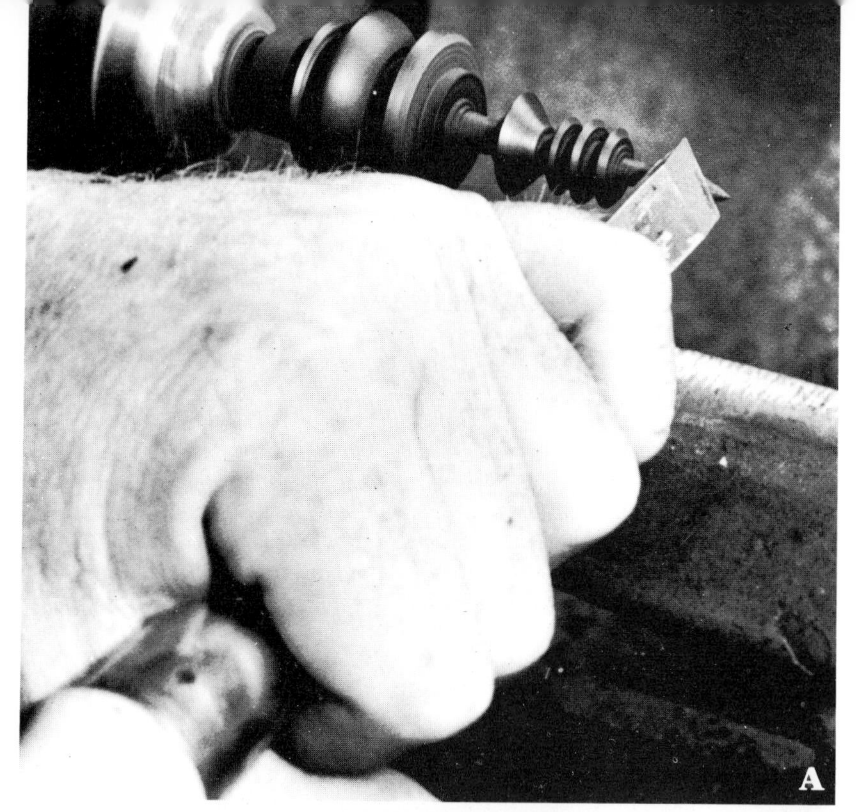

*A light shearing cut with the skew (**A**) creates a spiral flute pattern on the pointed finial (**B**).*

*The edge and corner of a parting tool makes concentric circles of chatterwork (**C**).*

*A skew decorates the spherical barrel of the stopper (**D**).*

*Burnishing with a pointed stick of the same wood species as the spindle (**E**) polishes and highlights the pattern.*

From *Fine Woodworking* magazine (November 1984) 49:81-83

produced crude chatterwork on your own turnings. Here's how it happens: A slender spindle can flex as it spins, and tool pressure pushes it into an elliptical rather than circular path around its axis. At high speed, the flexing wood vibrates against the tool's edge, which leaves marks on the surface of the wood. Chatter can also be caused by a tool that's dull or held at the wrong angle, but this type of blemish hardly leaves a decorative surface.

I control chatter by carefully reducing the diameter of the spindle so it can flex easily, then I delicately manipulate the work with a razor-sharp tool to make it chatter. Chatterwork is workmanship of extreme risk—I'm practically daring the spindle to break, and often it obliges. Little explosions of exotic wood, pieces and splinters flying, are part and parcel of the technique. Needless to say, eye protection is essential.

I work on a small Duro lathe equipped with a ¾-in. Jacobs chuck on the headstock spindle. For most of my turning, I use a ½-in. skew, and a ½-in. wide parting tool ⅛ in. thick. I also make small parting tools out of ¼-in. square steel key stock (normally used for keying pulleys to motor shafts). I only use a tool for a few quick cuts before regrinding, so the key stock holds an edge long enough to suit me. My homemade parting tools are designed to cut a very small amount of material at a time, to minimize resistance between tool and work. I rarely use these special parting tools for the actual chattering procedure.

Nothing affects final results more than tool sharpness. I grind all my tools to one of two basic angles: 60° for parting tools and 75° for skews. I keep a two-wheel grinder next to the lathe, with the parting-tool angle set on one wheel and the skew angle on the other. Because I resharpen so often, I don't have time to reset tool rests. I prefer a fresh razor burr straight from the medium wheel, so I never hone or whet a tool by hand.

Good light is essential for proper reading of the chattered surface, so my lathe sits under a combination of fluorescent and incandescent lights, including a flex-arm desk lamp on the ways. Behind the lathe, I've tacked a Masonite panel, painted white to reflect light on the spindle.

Here's how I begin a stopper. From a 7-in. long, 1-in. square blank mounted between centers, I rough out four stoppers without cutting them apart. For a headstock center, I clamp a pointed ¼-in. dia. steel dowel in the chuck. With the tailstock tightened enough for the steel dowel to drive the spindle, I turn a ½-in. dia. tenon on the tailstock end of the piece. Then I flip the stock end-for-end, and clamp the tenon in the chuck. After roughing out four stoppers, I part through the spindle. This leaves one stopper attached to the tenon in the chuck, and the other three, each with its own tenon, on the rest of the blank, which I set aside.

Before removing too much material near the headstock, I turn the stopper to its final shape. Since there's no tailstock support, the piece is already somewhat flexible and beginning to chatter, particularly at the unsupported end. In fact, the form and the finished texture are occurring simultaneously because I'm often producing a changing progression of chattered surfaces as I'm turning the shape. Each moment, the changing texture of the piece suggests changes in my original design. Does a dramatic development in the texture merit a revision of form? When has each section of the turning reached its ideal state? There's an interesting tension as the work develops.

When I'm satisfied with the shape, I sand the areas that won't have decoration. I don't cover an entire piece with chatterwork—it looks best when contrasted next to a smooth surface. I sand as little as possible. Preserving the sharp edges of a turning gives a vitality that's lost in a heavily sanded piece.

Here I'll explain three chatterwork patterns I use on a stopper. Experience has taught me what patterns I'm likely to get with, say, the edge of a skew or the corner of a parting tool, but there are lots of variables and I'm often surprised by the results. Experimentation is the only way to learn what patterns are possible.

For chatterwork, I run the lathe between 1400 RPM and 2200 RPM. Since the stopper spindle is already very thin near the pointed finial, there is usually plenty of flex there. I make a light shearing cut with the edge of a sharp skew—just one quick pass—along the conical profile. This makes a nice spiraling flute pattern. If the tool bites too deeply, the work will explode into useless fragments. After each cut, I turn off the lathe and inspect the work. If the first attempt doesn't yield a pleasing surface, I try another cut. This time I'll force a slightly greater flex by applying light downward pressure on the spindle as the cut progresses. If I'm still not getting enough flex, I use a parting tool to reduce the diameter of the stopper slightly on the headstock side of the finial.

I decorate the stopper's disc with a series of concentric circles of radial flutes (or facets). With the edge and corner of a sharp parting tool, I begin at the center of the disc where the finial emerges, and lay in concentric circular ridges. Slight pressure toward the headstock deflects the spindle and initiates the chatter as I cut each circle. Again, each cut takes only a few seconds, and I stop the lathe after every cut or two to inspect the work. During these pauses I regrind the cutting edge—at least every fourth or fifth cut. I keep the ring-chatter cuts shallow, and always plan enough thickness to the disc to allow me to shear off the chattered surface and begin a second or even third series of chatter cuts if I'm not satisfied. Often I have to reduce the thickness of the spindle to give more flex. Fluted rings are fairly easy. You should be able to cut them on the first or second try.

On the spherical barrel, I make a pattern that resembles the texture of beaten metal. I get this effect by making a smooth cut with a sharp skew just as if I were reducing the diameter. Since the stem diameter was reduced after the shape was completed, the skew now chatters over the surface instead of reducing the diameter. The chatter is shallow, but noticeable. With one pass down the headstock side of the barrel and a mirror-image pass down the tailstock side, amazing patterns can emerge.

After I've chattered a surface, I burnish it before going on to the next one. This accentuates the texture, polishes the high points, and reveals any errors. My burnisher is simply a ¼-in. square stick, 12 in. to 18 in. long, of the same species as the spindle being burnished. I sharpen each end to a point. (The ends are quickly blunted, so it's handy to have several burnishers sharpened in advance before each turning session.) One pass over the spinning chatterwork does it.

When I'm finished, I examine the piece for flaws, moving the light around the work to highlight the surface. Sometimes light sanding is necessary. Then I apply Watco with cheesecloth or other lint-free fabric while the stopper is still on the lathe, and I polish with a clean, dry cloth. I part the piece at the stem and set it aside. When the finish is dry, I buff the stopper on a linen wheel or wool bonnet lightly dressed with pure carnauba wax. □

Stephen Paulsen earns his living producing scent bottles, jewelry boxes and small containers in Goleta, Calif.

Index

FINE WOODWORKING
Editorial Staff, 1975-1986

Paul Bertorelli
Mary Blaylock
Dick Burrows
Jim Cummins
Katie de Koster
Ruth Dobsevage
Tage Frid
Roger Holmes
Cindy Howard
John Kelsey
Linda Kirk
Nancy-Lou Knapp
John Lively
Rick Mastelli
Nina Perry
Jim Richey
Paul Roman
David Sloan
Nancy Stabile
Laura Tringali
Linda D. Whipkey

FINE WOODWORKING
Art Staff, 1975-1986

Roger Barnes
Kathleen Creston
Deborah Fillion
Lee Hov
Betsy Levine
Lisa Long
E. Marino III
Karen Pease
Roland Wolf

FINE WOODWORKING
Production Staff, 1975-1986

Claudia Applegate
Barbara Bahr
Jennifer Bennett
Pat Byers
Mark Coleman
Deborah Cooper
Kathleen Davis
David DeFeo
Michelle Fryman
Mary Galpin
Dinah George
Barbara Hannah
Annette Hilty
Margot Knorr
Jenny Long
Johnette Luxeder
Gary Mancini
Laura Martin
Mary Eileen McCarthy
JoAnn Muir
Cynthia Lee Nyitray
Kathryn Olsen
Mary Ann Snieckus
Barbara Snyder